Smart Innovation, Systems and Technologies

Volume 236

D1800004

Series Editors

Robert J. Howlett, Bournemouth University and KES International,
Shoreham-by-Sea, UK

Lakhmi C. Jain, KES International, Shoreham-by-Sea, UK

The Smart Innovation, Systems and Technologies book series encompasses the topics of knowledge, intelligence, innovation and sustainability. The aim of the series is to make available a platform for the publication of books on all aspects of single and multi-disciplinary research on these themes in order to make the latest results available in a readily-accessible form. Volumes on interdisciplinary research combining two or more of these areas is particularly sought.

The series covers systems and paradigms that employ knowledge and intelligence in a broad sense. Its scope is systems having embedded knowledge and intelligence, which may be applied to the solution of world problems in industry, the environment and the community. It also focusses on the knowledge-transfer methodologies and innovation strategies employed to make this happen effectively. The combination of intelligent systems tools and a broad range of applications introduces a need for a synergy of disciplines from science, technology, business and the humanities. The series will include conference proceedings, edited collections, monographs, handbooks, reference books, and other relevant types of book in areas of science and technology where smart systems and technologies can offer innovative solutions.

High quality content is an essential feature for all book proposals accepted for the series. It is expected that editors of all accepted volumes will ensure that contributions are subjected to an appropriate level of reviewing process and adhere to KES quality principles.

Indexed by SCOPUS, EI Compendex, INSPEC, WTI Frankfurt eG, zbMATH, Japanese Science and Technology Agency (JST), SCImago, DBLP.

All books published in the series are submitted for consideration in Web of Science.

More information about this series at http://www.springer.com/series/8767

Lakhmi C. Jain · Roumen Kountchev ·
Yonghang Tai
Editors

3D Imaging Technologies— Multidimensional Signal Processing and Deep Learning

Methods, Algorithms and Applications, Volume 2

 Springer

Editors
Lakhmi C. Jain
Centre for Artificial Intelligence
University of Technology Sydney
South Australia, NSW, Australia

Roumen Kountchev
Faculty of Telecommunications
Technical University of Sofia
Sofia, Bulgaria

Yonghang Tai
Yunnan Normal University
Kunming, China

ISSN 2190-3018 ISSN 2190-3026 (electronic)
Smart Innovation, Systems and Technologies
ISBN 978-981-16-3182-5 ISBN 978-981-16-3180-1 (eBook)
https://doi.org/10.1007/978-981-16-3180-1

This Springer imprint is published by the registered company Springer Nature Singapore Pte Ltd.
The registered company address is: 152 Beach Road, #21-01/04 Gateway East, Singapore 189721,
Singapore

Preface

This book contains the papers presented at the 2nd International Conference on 3D Imaging Technologies—Multidimensional Signal Processing and Deep Learning (3D IT-MSP&DL) which was carried out on December 11–13, 2020, at the Yunnan Normal University, Kunming, China. The papers are aimed at the contemporary areas of the 3D image representation, 3D image technology, 3D image and graphics, multidimensional signal, image, and video processing and coding, and the related mathematical approaches and applications. The advance of the contemporary computer systems for processing, analysis, and recognition of patterns and situations opens new abilities, beneficial to practice. As a result, a synergic combination of innovative theoretical investigations, approaches, and applications is achieved.

The book: 3D Imaging Technologies—Multidimensional Signal Processing and Deep Learning (3D IT-MSP&DL), comprises 2 volumes:

- *Mathematical Approaches and Applications (Volume 1);*
- *Methods, Algorithms, and Applications (Volume 2).*

The main topics of the chapters in volume 2 are:

Spectral reflectance reconstruction; drawing of digital printing gamut map; improved contour feature extraction for the image butterfly specimen; improved NN-Unet for surface defect segmentation; segmentation algorithm of Dongba scripture; flower gender recognition based on YOLO v4; feature understanding based on deconvolution visualization; resource demand prediction and optimal allocation in cloud computing environment; incentive effect of tax preference on enterprise innovation; social security helping rural poverty alleviation; sensitivity analysis of NVH structure of car body; design of nutrition and sports health management system; English word intelligent learning system based on mobile education concept; application of big data analysis in English language education; analysis of droplet diffusion in negative pressure ward; cognitive analysis of command mechanism based on operational data flow motifs; confirmatory factor analysis of place attachment in leisure tourism; application of computer sensor and detection technology in mechatronics; smart customer service system based on data mining; natural language processing service in intelligent customer service system; university network ecological governance based on block chain technology; Informatization Construction of University Laboratory under the background of new infrastructure; run time prediction practices of multimedia web design in technology management; simulation of orthogonal frequency division multiplexing signaling; intelligent financial framework of colleges and universities in information age; VHF band dual channel high-power transmitter and receiver module; implementation of point multiplication over elliptic curve cryptography; non-contact hand sanitizer auxiliary device; simulation of Young's double slit interference experiment; research of the ferroelectric memory using ferroelectric field-effect transistors; characteristics of molybdenum disulfide field-effect transistor; microcontroller-based coffee acidity detection design; measuring acceleration of gravity with simple pendulum; design of intelligent boiler controller; fire alarm based on single chip microcomputer; multi-input high fidelity preamplifier; the early warning and diagnosis information database of power plant network security events; prediction algorithm of atomic clock combination clock difference; smart humidifier based on voice control; image target detection using the YOLO v5 algorithm.

The aim of the book is to present the latest achievements of the authors, to a wide range of readers: IT specialists, engineers, physicians, Ph.D. students, and other specialists.

Acknowledgments The book editors express their special thanks to book chapter reviewers for their efforts and good will to help for the successful preparation of the book. Special thanks to Prof. Lakhmi C. Jain (Honorary Chair), Prof. Dr. Srikanta Patnaik, and Prof. Dr. Junsheng Shi (General Chairs) and Dr. Yonghang Tai (Organising Chair) of 3D IT-MSP&DL.

The editors express their warmest thanks to the excellent Springer team which made this book possible.

South Australia, Australia	Lakhmi C. Jain
Sofia, Bulgaria	Roumen Kountchev
Kunming, China	Yonghang Tai
April, 2021	

Contents

About the Editors

Lakhmi C. Jain, B.E. (Hons), M.E., Ph.D., Fellow (Engineers Australia), served as Visiting Professor in Bournemouth University, UK, until July 2018 and presently serving the University of Canberra, Australia and University of Technology Sydney, Australia. Dr. Jain founded the KES International for providing a professional community the opportunities for publications, knowledge exchange, cooperation and teaming. Involving around 5000 researchers drawn from universities and companies worldwide, KES facilitates international cooperation and generates synergy in teaching and research. KES regularly provides networking opportunities for professional community through one of the largest conferences of its kind in the area of KES.

Prof. Roumen Kountchev, Ph.D., D.Sc., Technical University of Sofia, Bulgaria. His scientific areas of interest are digital signal and image processing, image compression, multimedia watermarking, video communications, pattern recognition and neural networks. He has 341 papers published in magazines and conference proceedings (71 international); 15 books; 46 book chapters; 20 patents (3 international). He had been Principle Investigator of 38 research projects (6 international). He is Member of Euro Mediterranean Academy of Arts and Sciences (EMAAS), President of Bulgarian Association for Pattern Recognition (Member of Intern. Association for Pattern Recognition), Editorial Board Member of IJBST Journal Group, Member of International Research Institute for Economics and Management (IRIEM), Member of Institute of Data Science and Artificial Intelligence (IDSAI), Member of the Honorable Editorial Board of the nonprofit peer-reviewed open access IJBST Journal Group. Editorial Board Member of International Journal of Reasoning-based Intelligent Systems; International Journal Broad Research in Artificial Intelligence and Neuroscience; KES Focus Group on Intelligent Decision Technologies; Egyptian Computer Science Journal; International Journal of Bio-Medical Informatics and e-Health; and International Journal Intelligent Decision Technologies.

Prof. Yonghang Tai is a professor at the School of Physics and Electronic Information, Yunnan Normal University, Kunming, China, Color and Image Vision Lab. He got his M.Sc. at Yunnan Normal University, Kunming, and Ph.D. in Opto-Electronic Engineering (OE) at Deakin University, Melbourne, Australia. His main research interests are in 3D HMD design, AM-OLED drive circuit design, and Stereoscopic imaging systems. Prof. Tai had published many journal and conference papers and has 6 patents. He took part in the Electronic Nose Designing of Tobacco Mildew Early Warning Project and Designed and debugged the AM-OLED Gamma correction system of Yunnan North OLEiD Company. Prof. Tai was the editor and reviewer of many indexed journals. He was a tutor of M.Sc. and Ph.D. students who successfully defended their works.

Spectral Reflectance Reconstruction Based on Weighted Root Polynomial

Mingjiang Ji, Junsheng Shi, Yonghang Tai, and Wanli Jiang

Abstract The reconstruction of spectral reflectance by the RGB values of the camera response can not only improve the efficiency of spectral reflectance reconstruction but also reduce the reconstruction cost. However, the accuracy of spectral reflectance reconstruction is not high only through the three-channel response value. The accuracy of spectral reflectance reconstruction is also related to camera exposure. In this paper, a method of a weighted root polynomial is researched using Canon EOS 500D, Digital Color Checker SG as the training samples, Color Checker Rendition Chart as the test samples, and root-mean-square error (RMSE) and goodness of fit coefficient (GFC) as the evaluation criteria. According to the experimental results, the method can maintain the reconstruction accuracy under the same conditions and reduce the impact of the exposure changes on reconstruction accuracy when the camera exposure time changes.

1 Introduction

Spectral reflectance is usually referred to as the "fingerprint" of an object [1]. The spectral reflectance of different objects' surface colors is different. Spectral reflectance is an objective attribute of an object and does not vary with the influence of observation conditions and other factors, so it is suitable for the transmission of color information and high-fidelity reproduction. The spectral reflectance of the object can be obtained through the surrounding environmental information, light conditions, the output image, and the camera's spectral sensitivity. According

M. Ji · J. Shi (✉) · Y. Tai (✉) · W. Jiang
School of Physics and Electronic Information, Yunnan Normal University, Kunming, China
e-mail: shijs@ynnu.edu.cn

Y. Tai
e-mail: taiyonghang@126.com

Yunnan Key Lab of Optic-Electronic Information Technology, Kunming, China

to different reconstruction principles, the existing spectral reflectance reconstruction methods can be divided into four categories: interpolation reconstruction, direct reconstruction, learn-based reconstruction, and combined reconstruction [2–4].

The traditional spectral reflectance measurement is done by spectrophotometer, but the general spectrophotometer can be measured in a small range and can only measure the plane, and the work efficiency is low. A multi-spectral camera can obtain the spectral reflectance of each point of the object, but it is expensive and inconvenient to carry, which limits its application in reality. At present, some researchers collect the color information of the object surface through a digital camera, the RGB value of the object color can be obtained, and then, the spectral reflectance can be calculated from the RGB value through conversion matrix. This method of spectral reflectance obtained through conversion is called spectral reflectance reconstruction. A spectral reflectance reconstruction algorithm is used to reconstruct the spectral reflectance of each pixel, and finally, the original image is reproduced by computer image graphics technology [5]. However, there is an inevitable correlation between the color information acquired by the digital camera and the equipment, and the number of channels is only three, so the accuracy of spectral reflectance reconstruction obtained from this information is not high. Obtaining the spectral reflectance is an important part of achieving spectral color reproduction. Therefore, how to obtain high-precision spectral reflectance is very important. It has become a hot spot in the field of color science research. It has broad application prospects in application fields that require high reconstruction accuracies, such as color reproduction, color printing, and restoration and reproduction of artworks [6].

2   Model and Method

A schematic diagram of an imaging system and imaging process is shown in Fig. 1. $I(\lambda)$ represents the spectrum of the light source, $r(\lambda)$ represents the spectral reflectance, $t_k(\lambda)$ (k=R,G,B) represents the spectral transmittance associated with

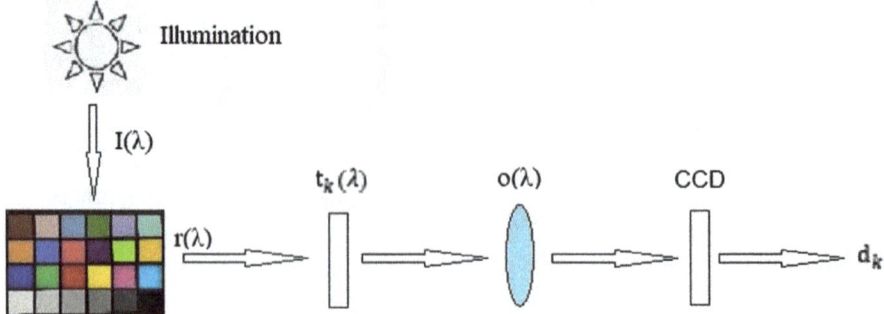

Fig. 1 Schematic diagram of an imaging system and imaging process

the kth filter, $o(\lambda)$ represents the spectral transmittance of the optical system, $c(\lambda)$ represents the spectral sensitivity of the CCD, and η_k (k=R,G,B) is the spectral noise for the kth channel. The camera output d_k (k=R,G,B) is related to the channel k for a single pixel or region in the image.

$$d_k = \int_{\lambda_{\min}}^{\lambda_{\max}} I(\lambda) r(\lambda) o(\lambda) c(\lambda) t_k(\lambda) d\lambda + \eta_k \tag{1}$$

Assume that the noise has been removed by preprocessing. We can replace $c(\lambda){\cdot}o(\lambda){\cdot}t_k(\lambda)$ with the spectral sensitivity $S_k(\lambda)$ (k=R,G,B) of the kth channel. Then, Eq. (1) can be written as

$$d_k = \int_{\lambda_{\min}}^{\lambda_{\max}} I(\lambda) r(\lambda) S_k(\lambda) d\lambda \tag{2}$$

Then, Eq. (2) can be expressed as a matrix

$$d_k = I S_k r \tag{3}$$

Considering all channels of this system, Eq. (3) can be expressed as

$$d = I S r \tag{4}$$

where d is a matrix containing all the camera output d_k, S is a matrix containing all the channel spectral sensitivity S_k. The ultimate goal is to restore $r(\lambda)$ from the camera output according to Eq. (4). This is done by finding a transformation matrix Q, which has the following solution:

$$r = Q d \tag{5}$$

Research shows that it is difficult to obtain ideal reconstruction accuracy when using three original channels of the color digital camera for spectral reconstruction [7]. Liang et al. showed that the spectral reflectance reconstruction method extended by camera response value and local inverse distance weighted optimization can effectively improve the reconstruction accuracy [8]. Graham et al. showed that for a given standard training set, a polynomial color correction could significantly reduce colorimetric error [9]. However, polynomial color correction fitting is dependent on exposure, namely, with the change of exposure, the polynomial component of the vector changes in a nonlinear manner, leading to changes of hue and saturation. So, they propose a new method which they call "Root-Polynomial Color Correction" (RPCC) [9]. Experimental results show that RPCC enhances the color correction performance of real data and synthetic data.

Based on the above analysis, a weighted square root polynomial regression algorithm is proposed. For the learning reconstruction method, the higher the spectral and chromaticity attribute similarity between the test sample and the training sample, the higher the reconstruction accuracy, and the lower the similarity, the lower the reconstruction accuracy. In this study, the anti-Euclidean distance weighting method is adopted, that is, the reciprocal of the distance between the test sample and the $j(1 \leq j \leq N)$ training sample is the ω_j [10]:

$$\omega_j = \frac{1}{\sqrt{\left(R_{\text{test}} - R_j\right)^2 + \left(G_{\text{test}} - G_j\right)^2 + \left(B_{\text{test}} - B_j\right)^2 + \sigma}} \tag{6}$$

where R_{test}, G_{test}, and B_{test} are the RGB three-way channel response values of the test sample. R_j, G_j, and B_j are the RGB three-way channel response values of the jth training sample. σ is a minimum number to prevent the denominator of Eq. (6) from being zero. The weights of any test sample and all training samples are reconstructed into a diagonal matrix W [11]:

$$W = \begin{bmatrix} \omega_1 & & & \\ & \omega_1 & & \\ & & \ddots & \\ & & & \omega_j \end{bmatrix}_{N \times N} \tag{7}$$

The fourth-degree root polynomial expansion is shown in Eq. (8):

$$d_{\text{expanded}} = (R, G, B, \sqrt{RG}, \sqrt{RB}, \sqrt{GB}, \sqrt[3]{R^2 G}, \sqrt[3]{R^2 B}, \sqrt[3]{G^2 B}, \sqrt[3]{G^2 R}, \sqrt[3]{B^2 R},$$
$$\sqrt[3]{B^2 G}, \sqrt[3]{RGB}, \sqrt[4]{R^3 G}, \sqrt[4]{R^3 B}, \sqrt[4]{G^3 B}, \sqrt[4]{G^3 R}, \sqrt[4]{B^3 R}, \sqrt[4]{B^3 G}, \sqrt[4]{R^2 GB},$$
$$\sqrt[4]{G^2 RB}, \sqrt[4]{B^2 GR})^{\text{T}} \tag{8}$$

According to Eq. (5), the spectral reflectance r of the test sample is

$$r = R_p W D_p^T \left(D_p W D_p^T\right)^{-1} d_{\text{expanded}} \tag{9}$$

where R_p represents the spectral reflectivity matrix of the training sample, D_p represents the extended response matrix of the training sample, W is the weight matrix, and d_{expanded} represents the extended response matrix of the test sample.

3   Experiments

3.1   Experimental Data Acquisition

This experiment uses Digital Color Checker SG as the training samples, Color Checker Rendition Chart as the test samples, and Canon EOS 500D. The PR-715 was used to measure the spectral information of the training sample and the test sample within the wavelength range of 400–700 nm. The device adopted the measurement condition of 45°/0°. Use digital camera Canon EOS 500D to capture photos. Some camera parameters are set as follows: ISO 100, aperture $f/5.6$, and exposure time is 1/10 and 1/25 s.

3.2 Evaluation Standard

Two indexes were used to evaluate and compare the accuracy of spectral reconstruction by different methods. The first index is goodness of fit coefficient, and the second index is root-mean-square error.

Goodness of fit coefficient (GFC):

$$\text{GFC} = \frac{\sum_{\lambda=\min}^{\max} R(\lambda) R'(\lambda)}{\sqrt{\sum_{\lambda=\min}^{\max} R(\lambda^2)} \sqrt{\sum_{\lambda=\min}^{\max} R'(\lambda^2)}} \tag{10}$$

For evaluation, fitness coefficient is a relatively new evaluation method, which is used to evaluate the benefits of spectral reflectance similarity without being affected by scale factors [12]. The value range of GFC is between 0 and 1. When the value of GFC is 1, the reconstruction effect is the best; when the value of GFC is 0, the reconstruction effect is the worst.

Root-mean-square error (RMSE):

$$\text{RMSE} = \sqrt{\frac{\sum_\lambda (R(\lambda) - (R'(\lambda))^2}{N}} \tag{11}$$

$R(\lambda)$ is the original spectral value, and $R'(\lambda)$ is the most commonly used error comparison method for reconstructed spectral values, with the smaller the value, the smaller the difference.

3.3 Experimental Results and Discussion

In this part, we use two experiments to prove the effectiveness of our method. In the first experiment, we collected training samples and test sample data under all conditions being the same. In the second experiment, the exposure time is set to 1/25 s when collecting training samples, and the exposure time is set to 1/10 s when collecting test samples. Three methods are used in the experiment. Method 1: Spectral reflectance is reconstructed by only three values of RGB; Method 2: Spectral reflectance was reconstructed by the fourth-order weighted polynomial regression

Table 1 Results of experiment 1

	GFC			RMSE		
	Mean	Max	Min	Mean	Max	Min
Method 1	0.9842	0.9972	0.9038	0.1210	0.3228	0.0604
Method 2	0.9926	0.9992	0.9507	0.0401	0.0714	0.0134
Our method	0.9909	0.9992	0.9461	0.0528	0.0959	0.0197

Table 2 Results of experiment 2

	GFC			RMSE		
	Mean	Max	Min	Mean	Max	Min
Method 1	0.9814	0.9975	0.9063	0.2847	0.3906	0.0980
Method 2	0.9465	0.9991	0.5479	0.4241	0.9050	0.0775
Our method	0.9868	0.9990	0.9218	0.2884	0.5441	0.0872

algorithm. Method 3: The spectral reflectance was reconstructed by the fourth-order weighted root polynomial regression algorithm.

Table 1 is the results of experiment 1:

From the data in Table 1, we can see that compared with method 1, the fitting effect of method 2 and the method in this research is better, the RMSE is significantly reduced, and the method of expanding camera response can significantly improve the accuracy of the reconstructed spectrum. And compared with the method in this research, the second method is the fourth-order extension, the number of extensions in this paper is less, and the calculation amount is smaller if the same precision can be achieved.

Table 2 is the results of experiment 2:

From the data in Table 2, we can see that the reconstruction effect after the change of exposure time is worse than that in experiment 1. But, the method proposed in this research is the one with the best effect. The researched method can effectively weaken the influence of the change of exposure time on the reconstruction accuracy.

4   Conclusion

In this study, a spectral reflectance reconstruction method based on weighted square polynomials is researched, and the reconstruction results are evaluated by GFC and ERMS. The researched method is compared with two existing classical methods through specific experiments. Experimental results show that the researched method can maintain the reconstruction accuracy under the same conditions and reduce the impact of the exposure changes on reconstruction accuracy when the camera exposure time changes.

Acknowledgements This work is funded by grants from the National Science Foundation of China (grant number 61650401, 61875171) and the Yunnan Education Commission of China (grant number ZD2014004).

References

1. Hajipour, A., Shams-Nateri, A.: Effect of classification by competitive neural network on reconstruction of reflectance spectra using principal component analysis. Color Res. Appl. **42**(2), 182–188 (2017)
2. Wang, W., Wang, J.: Research progress of spectral reflectance reconstruction technology. Packag. Eng. **41**(11), 254–261 (2020)
3. Yang, P., Liao, N.F., Song, H.: Study of approaches to spectral reflectance reconstruction based on digital camera. Spectrosc. Spectral Anal. **29**(5), 1176–1180 (2009)
4. Farhad, A., et al.: Reconstruction of reflectance data using an interpolation technique. J. Opt. Soc. Am. A **26**(3), 613–624 (2009)
5. Wang, H.W., Li, J., Chen, G.X.: Study on key issues of multi-spectral color reproduction technique. Adv. Mater. Res. **174**, 93–96 (2011)
6. Wang, H., Li, J., Wan, X., Gan, C.: Research on the printing color reproduction technology based on spectral imaging. Packag. Eng. **029**(004), 40–42 (2008)
7. Mansouri, A., Sliwa, T., Hardeberg, J.Y., Voisin, Y.: Representation and estimation of spectral reflectances using projection on PCA and wavelet bases. Color. Res. Appl. **33**(6), 485–493 (2008)
8. Liang, J., Wan, X.: Spectral reconstruction from single rgb image of trichromatic digital camera. Acta Opt. Sin. **37**(09), 370–377 (2017)
9. Finlayson, G.D., et al.: Color correction using root-polynomial regression. IEEE Trans. Image Process. A Publ. IEEE Sig. Process. Soc. **24**, 1460–1470 (2015)
10. Liang, J., Wan, X.: Optimized method for spectral reflectance reconstruction from camera responses. Opt. Exp. **25**(23), 28273 (2017)
11. Amiri, M.M., Fairchild, M.D.: A strategy toward spectral and colorimetric color reproduction using ordinary digital cameras. Color. Res. Appl. **43**(5), 675–684 (2018)
12. Amiri, M.M., Amirshahi, S.H.: A step by step recovery of spectral data from colorimetric information. J. Opt. **44**(4), 373–383 (2015)

A Drawing of Digital Printing Gamut Map

Dongdong Yang, Zaiqing Chen, Junsheng Shi, and Yonghang Tai

Abstract Based on the digital printing gamut map, the (Commission Internationale de L'Eclairage (CIE) data of different papers are read under the same international standard FOGRA39. In this paper, we used MATLAB programming to make visual three-dimensional color gamut map, analyze and summarize the gamut range of different paper. The color gamut map of 3D convex hull surface is drawn by algorithm, and the color gamut range of different paper under the same international standard FOGRA39 is realized. Through the data query and analysis on the drawn maps, the lab range of different paper is summarized.

1 Introduction

With the rapid development of science and technology, as one of the terminal media of information display, display has been gradually integrated into people's daily life, and people are pursuing a larger range of color gamut and higher definition. Naturally, three-dimensional gamut has become a research hotspot [1]. The gamut platform designed and developed in this paper can process 2D and 3D gamut data quickly and simply complete the conversion of color space, the calculation of 3D gamut volume and gamut coverage, and can save these data selectively. Researchers can evaluate the display through these data analysis and provide reference for understanding the display performance [2]. The operation of the platform is simple and the effect is intuitive. It is not only suitable for professional applications, but also can provide help for non-professionals [3]. The platform development environment is MTALABr2014a, and the operating system is Microsoft Windows7 [4].

D. Yang · Z. Chen (✉)
School of Information Science and Technology, Yunnan Normal University, Kunming, Yunnan, China
e-mail: Zaiqingchen@ynnu.edu.cn

D. Yang · Z. Chen · J. Shi · Y. Tai (✉)
Yunnan Key Laboratory of Optoelectronic Information Technology, Kunming, Yunnan, China
e-mail: taiyonghang@126.com

© The Author(s), under exclusive license to Springer Nature Singapore Pte Ltd. 2021
L. C. Jain et al. (eds.), *3D Imaging Technologies—Multidimensional Signal Processing and Deep Learning*, Smart Innovation, Systems and Technologies 236,
https://doi.org/10.1007/978-981-16-3180-1_2

2   Method

2.1   Main Contents of the Project

Based on the above background analysis of the basic factors causing the above problems [5], the CIE data of different papers printed under the same international standard FOGRA39 were measured by experiments. Through the conversion formula of color space lab-xyz-rgb [6], the corresponding algorithm was constructed, and the 3D visualization 3D gamut map was made by using the relevant mapping software MATLAB, and the scope of the 3D gamut map was studied. For example, the lab range of CIE data of different paper is analyzed and summarized, and the volume range of corresponding renderings is calculated [7]. Finally, the conclusion is a drawn map, which provides a reference for the color gamut range of the display.

2.2   Purpose of the Experiment

The standard lab value of FOGRA39 is obtained. At the same time, the calibrated digital ink jet printer Epson 9080 is tested. In order to obtain the lab values of all color blocks in the ECI2002 chart of five kinds of paper, the color gamut diagram of MATLAB is drawn to get the color gamut comparison between standard color gamut and different paper. The color gamut gap of paper in the same standard printer is analyzed, and the color tube direction is defined.

2.3   Experimental Equipment

ECI2002 chart is a set of input data developed by European color initiative (ECI) to describe the characteristics of four color printing.

Software: ProfileMaker is an excellent color management software, in which profile editor is used for standard International Color Consortium (ICC) conversion lab, and measure tool is used for ECI2002 chart measurement to obtain chart color block lab of different paper.

Equipment: EyeOne Isis is a professional color measurement instrument especially for measuring a large number of charts or creating a large number of ICC test charts.

2.4 Experimental Steps

1. The international standard ICC coated FOGRA39 is converted into standard lab value by using profile editor. The absolute conversion is usually suitable for digital sampling, which is used as a standard for comparison.
2. Taking ICC coated FOGRA39 as the adjustment curve, the calibrated Epson 9080 printer is used to print the ECI2002 charts of five kinds of paper, including Xuan paper, smooth art paper, rough art paper, pantek rough paper, and bright light paper.
3. The software measure tool and EyeOne Isis were used to read out ECI2002 charts on different paper, and the lab values of all color blocks in all charts were obtained as the gamut drawing data of MATLAB.
4. The color gamut is drawn and compared by MATLAB, and the experimental results are obtained.

2.5 Data Processing and 3D Convex Hull Surface Graph

In the drawing of 3D convex hull surface graph, a, b, L are stored in a matrix variable U (a, b, L). A group of triangular convex hull algorithm information can be established and saved in the matrix variables by using the convhulln function provided in MATLAB to call u (a, b, L) matrix [8], Then, through trisurf function, the information data of the convex hull algorithm of the triangular patch are stored in a, B, and L coordinate system, but because trisurf is random in coloring each convex hull surface, it is necessary to fill the corresponding color for each convex hull surface by using the patch function in MATLAB. Before using the patch function to fill the color, it is necessary to call out the order of each vertex (vertex order of triangular patch) to find the L, a, b, c, m, y, k data of each vertex and use it. The previous formula calculates the corresponding R, G, B and saves them in the color matrix C. Finally, we can draw the 3D color map by using fac, LPA, LPA, and LPA, where 1 represents the color transparency of the patch [9]. By using [Lim, v] = convhulln (U) statement, the volume of 3D convex hull (3D gamut) can be calculated directly, and its value can be stored in variable V to provide data support for later comparison.

3 Results and Discussion

Figures 1, 2, 3, 4, 5, and 6 show the color gamut of standard FOGRA39 and three-dimensional convex surface color gamut of different paper.

The data are obtained by drawing gamut map and calculating color and volume of six kinds of paper as listed in Table 1.

According to the above data, it can be concluded that the color gamut range of five kinds of paper is in the order of small to large [10].

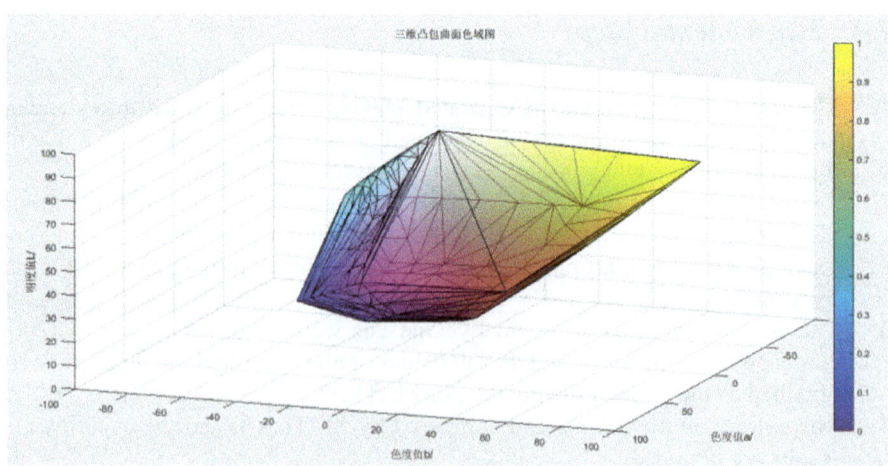

Fig. 1 Color gamut map of 3D convex hull surface

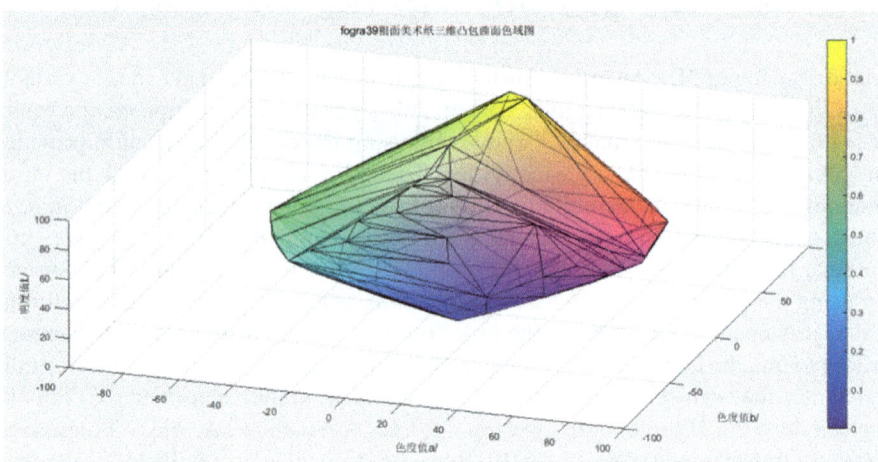

Fig. 2 Gamut map of 3D convex hull surface of FOGRA39 rough art paper

FOGRA39 rice paper < FOGRA39 smooth art paper < FOGRA39 fantec rough paper < FOGRA39 rough art paper < FOGRA39 high gloss.

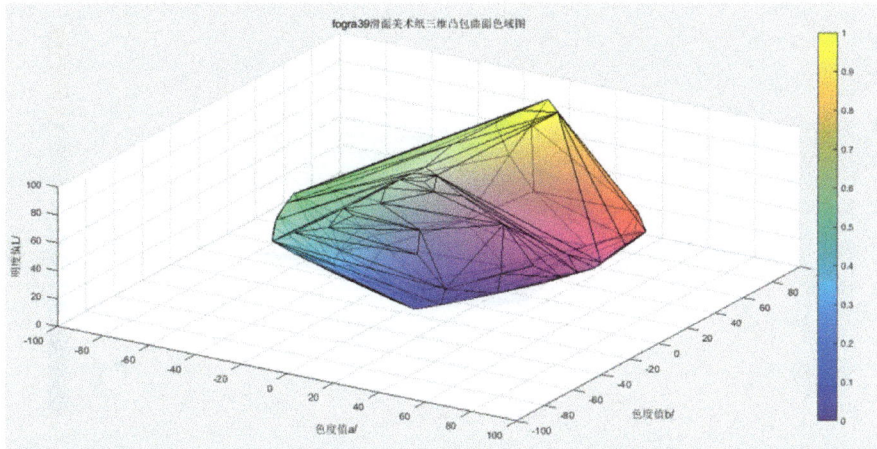

Fig. 3 Gamut map of 3D convex hull surface of FOGRA39 paper

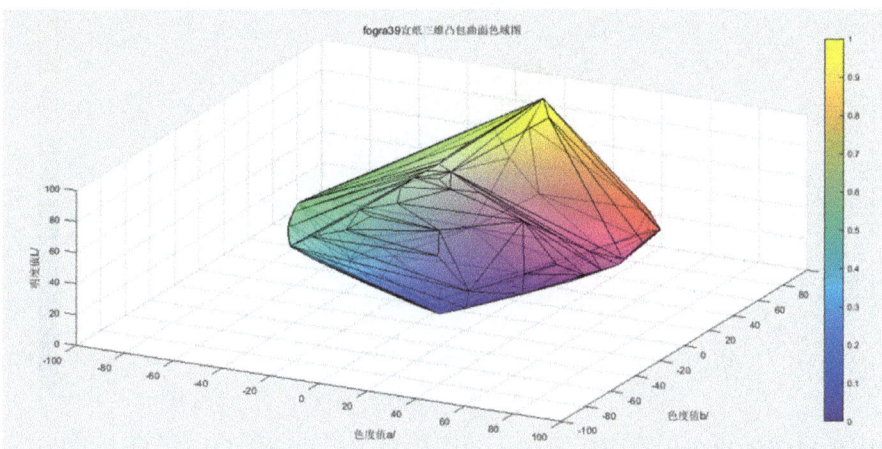

Fig. 4 Gamut map of three-dimensional convex hull surface of FOGRA39 rice paper

4 Conclusions

Based on the theory of digital printing and gamut map, this paper constructs the corresponding algorithm and realizes the drawing of three-dimensional visual gamut map through MATLAB platform. It laid a foundation for the later research on gamut.

In this paper, we obtained the standard data of FOGRA39 and CIE data of five different kinds of paper under FOGRA39 standard by experiment, which are FOGRA39 coarse art paper, FOGRA 39 smooth art paper, FOGRA39 fantec rough paper, FOGRA39 high light, and FOGRA39 Xuan paper. The three-dimensional

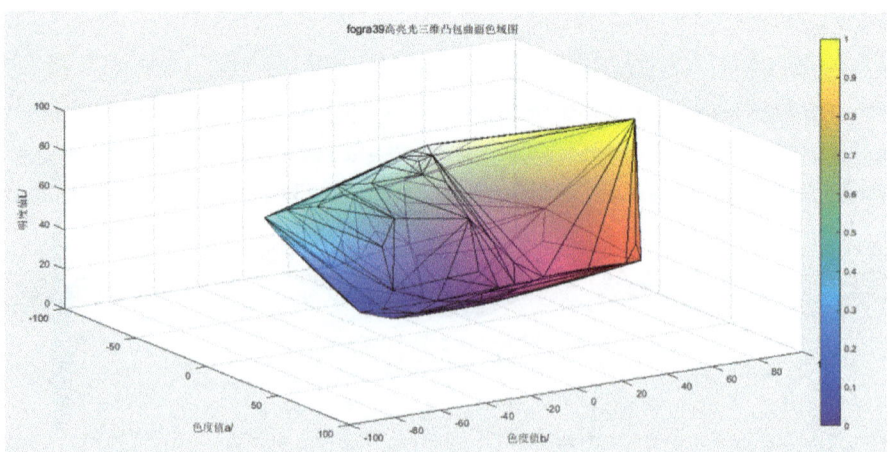

Fig. 5 Color gamut map of FOGRA39 high light 3D convex hull surface

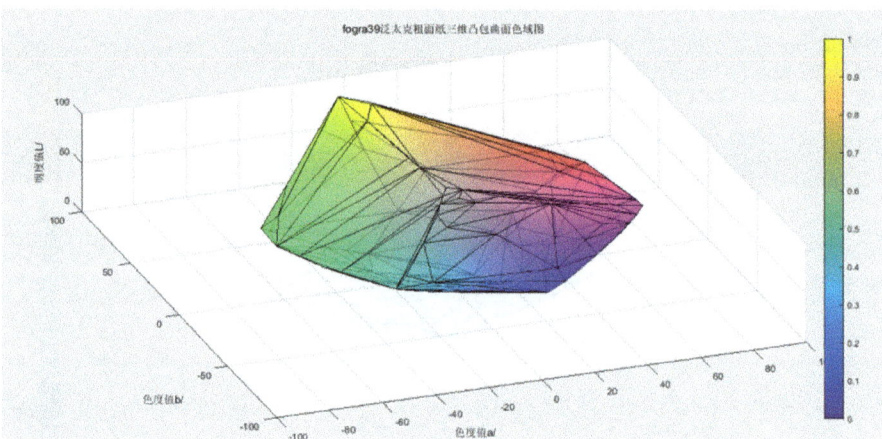

Fig. 6 FOGRA 39 fantec

Table 1 Gamut volume of various paper

Paper	FOGRA39 standard	FOGRA39 rough art paper	FOGRA39 fantec
Gamut volume	423,295	439,706	432,420
paper	FOGRA 39 high light	FOGRA39 glossy art paper	FOGRA 39 rice paper
Gamut volume	440,951	432,347	423,212

gamut map of each kind of paper is generated by the debugging program, and the volume of each paper's three-dimensional gamut and the gamut area of L-layer cross section are calculated and compared.

Acknowledgements This work was supported in part by grants from the National Natural Science Foundation of China (grant number: 61865015, 61875171, 61741516) and Key Program of the Yunnan Education Commission of China (grant number: 2018FA033).

References

1. Song, C., Wang, R.G., Chen, Y.: Fast calculation of color gamut boundary of LED display screen. Acta luminescens Sin. **07**, 31–35 (2013)
2. Hu, D.F., Zhang, Q.: Preliminary study on gamut matching method in color management. Printing Qual. Standard. **10**, 48–52 (2010)
3. Liu, Q.E.: Gamut and gamut coverage. Television Technol. **11**, 8–22 (2004)
4. Li, Y.: Design and development of gamut platform based on Matlab GUI. J. Tianjin Normal Univ. **03**, 24–30 (2014)
5. Wang, W., Hou, W., Chen, G.D.: Study on gamut boundary based on radial basis function interpolation. Sci. Technol. Inf. **16**, 16–18 (2011)
6. Yang, L., Liu, Z.: Application of least square method with interpolation in printer gamut boundary description. Packag. Eng. **11**, 14–18 (2013)
7. Li, N.: Study on data processing and theoretical construction of three dimensional gamut. J. Tianjin Normal Univ. **01**, 11–18 (2003)
8. Wang, K., Gao, Y.L., Lin, Y.: 3D color gamut measurement of LCD display and comparison of different color space applications. Optoelectron. Technol. Appl. **06,** 18–22 (2011)
9. Lin, M.H., Zhou, S.S., Luo, Y.H.: Image gamut visualization based on rotating ball algorithm. China Printing Packag. Res. **03**, 39–44 (2010)
10. Li, D., Zong, H.C., Xu, L.P.: Automatic selection of color gamut mapping intention based on ICC Standard. Packag. Eng. **12**, 16–18 (2009)

An Improved Contour Feature Extraction Method for the Image Butterfly Specimen

Fan Li and Wenjin Zhou

Abstract Automatic recognition of butterfly species is an important and effective method to develop and utilize butterfly resources. It is beneficial not only for the timely prevention and treatment of pests in agriculture and forestry, but also for the identification of rare butterfly species in the world. At present, some specific insect species have been well differentiated in some researches. However, due to the lack of rationality of classification feature design (which fails to correlated to the biological characteristics of butterflies effectively), there are still problems of poor generalization ability, low recognition rate, and so on. Therefore, in this paper, we discussed the effects of different types of features on butterfly classification problems and improve the histograms of multiscale curvature (HoMSC) calculation method to extract the shape features of butterfly wings. To prove the effectiveness of this method, we use a weight-based k-nearest neighbor (KNN) search algorithm and 400 images of 20 butterfly species (which belong to six different families) for testing. The accuracy rate of this method reached 96%. The result suggested the improved HoMSC features can be efficient for the identification of butterfly species which are closely related.

1 Introduction

Butterfly (Rhopalocera) is the second largest species in insects, which has a rich biodiversity [1]. In addition, butterfly has high values in aesthetics, economy, and ecology. So a reliable species classification method is essential for the rational use of different species of butterflies. Traditional methods and biochemical methods as the conventional solutions have been widely used to solve the problem of butterfly classification and recognition. However, these two types of methods need the professionals and the specialized equipment. Meanwhile, they have other problems, such as high cost, time consumption, and so on. Consequently, to realize the precision,

F. Li (✉) · W. Zhou
Kunming University of Science and Technology, Kunming 650500, China

© The Author(s), under exclusive license to Springer Nature Singapore Pte Ltd. 2021   17
L. C. Jain et al. (eds.), *3D Imaging Technologies—Multidimensional Signal Processing and Deep Learning*, Smart Innovation, Systems and Technologies 236,
https://doi.org/10.1007/978-981-16-3180-1_3

high efficiency, and automation of butterfly classification, a new method should be implemented [2].

In recent years, with the rapid development of artificial intelligence technology (AIT), more and more pattern recognition methods have been studied and applied to solve some practical problems (such as fingerprint recognition, face recognition, speech recognition, character recognition). Butterfly image as the most accessible pattern has been used for the research of automatic species recognition of butterfly in these years. Digital automatic identification system (DAISY) is a famous automated species identification system, which optimized for the rapid screening of invertebrates by non-experts. With the continuous improvement of the system from the mid-1990s to now, the recognition accuracy of 57 species of butterflies in the UK has been reached 98%. In addition, DAISY has been used in several research projects for solving different classification objects [3]. Except for mature system like this, some new methods have been developed to solve the problem of automatic identification of butterflies or other winged insects. In 2012, Jiangning Wang and Congtian Lin et al. designed an automatic identification to identify insect specimen images. In their research, several relative features were designed with the theory of taxonomy, and some common classifiers (like support vector machine and artificial neural networks) were used as pattern recognition methods [4]. In 2013, Yılmaz Kaya, Lokman Kayci presented a new method which is based on artificial neural networks (ANN), five textures and three color features for identification of butterfly species. In the result, they realized 92.5% recognition accuracy of 14 species of butterflies [5]. A new automatic identification system using photographic images had been designed to recognize fish, plant, and butterfly species from Europe and South America with Andrés Hernández-Serna and Luz Fernanda Jiménez-Segura. The automatic classification system integrates multiple image processing tools to extract the geometry, morphology, and texture of the images. Artificial neural networks (ANNs) were used as the pattern recognition method. The results show that the system performed with high accuracy, reaching 91.65% of true positive fish identifications, 92.87% of plants, and 93.25% of butterflies [6]. With the extensive application of deep learning technology in various fields in recent years, the research of automatic identification of butterfly images based on deep learning method was improved by Zhou Ai-Ming and Ma Peng-Peng et al. in 2017. By adjusting the Caffe framework, these researchers were able to identify 117 butterfly species of 6 families with 95.8% accuracy [7].

Although the above-mentioned methods are attractive and helpful for automatically identifying butterfly species, they still have some problems like complex, timeconsuming, and difficult to solve the image classification problem with high granularity. In addition, all the classification features, which are mentioned above, are common features of pattern recognition. Although these features have a good effect on the recognition of some objects, their discrimination ability for butterfly species with close relatives is limited. Therefore, we need to study more appropriate feature expression methods, which according to the characteristics of butterfly species to achieve more effective classification and recognition. In this paper, we improve the HoMSC calculation method (which presented in [8]) to extract the shape features of butterfly wings. On the one hand, the shape features are more stable than texture and color features (this conclusion will be elaborated in the following paragraphs). On the other hand, it is an efficient and accurate method to describe the concave–convex changes of butterfly wings.

2 Materials and Methods

2.1 Butterfly Species

In this paper, we used 400 images of 20 butterfly species (which belong to 6 family) for testing. These butterflies are: *Baronia brevicornis*, *Eurytides epidaus*, *Papilio machaon*, *Papilio pelaus*, *Chioides catillus*, *Myscelus amystis*, *Phocides polybius*, *Pyrrhopyge zenodorus*, *Danaus plexippus*, *Libytheana carinenta*, *Lycorea halia*, *Marpesia_chiron*, *Dismorphia eunoe*, *Kricogonia lyside*, *Pseudopierisnehemia nehemia*, *Hypaurotis crysalus*, *Leptotes cassius*, *Ithomiola floralis*, *Redrimmed euselasia*, and *Sarota subtesselata* (Fig. 1). All these images were downloaded from a Web site of butterfly specimens, and each of images was taken on top view and a white background. As shown in Fig. 1, these butterfly species are similar in whole or in part closed by their shapes.

Fig. 1. Twenty butterfly species used for automatic species identification

2.2 Feature Selection

Generally, three types of intuitive classification features can be easily obtained from digital images: color feature, texture feature, and shape feature, which often present different effects for different classification objects. In our previous research, we also tried to use them to extract features of butterfly specimen images, respectively, and combined with some common classification methods to carry out identification research [8, 9].

However, combined with our previous research results, we found that the color feature presented obvious differences in images of different types of butterfly specimens, but due to the influence of the shooting environment, specimen storage time, and other factors, the feature also changed among the same species of butterflies. In addition, the wing surface texture of some species of butterflies often changes greatly with the growing environment of butterflies and the collection season of specimens, which affects the effect of texture feature as a universal classification feature. By contrast, the shape of a butterfly's wing plane is a stable feature (although there are a few instances of morphological changes in the same species). When extracting shape features, the design of extraction method can avoid the instability caused by changes in shape size, shooting angle, shooting direction, and detail defects. Therefore, we improve the HoMSC calculation method for feature extraction according to the changes of the overall and local curvature of the wing surface contour of butterfly specimens.

2.3 The Process of Feature Extraction

In this paper, we organize the feature extraction process as Fig. 2 shows. In order to getting a clear contour feature, the image should be converted to binary graph. After that, for reducing the amount of computation (according to the symmetry of butterfly wings), we separate the left wing of butterfly for later feature extraction. Before the extraction of texture features, the image is blocked to two areas (the fore wing and the under wing). Finally, we extract the contour feature through the improved HoMSC calculation method from the two areas. The extraction steps and calculation methods of the HoMSC are described in detail below.

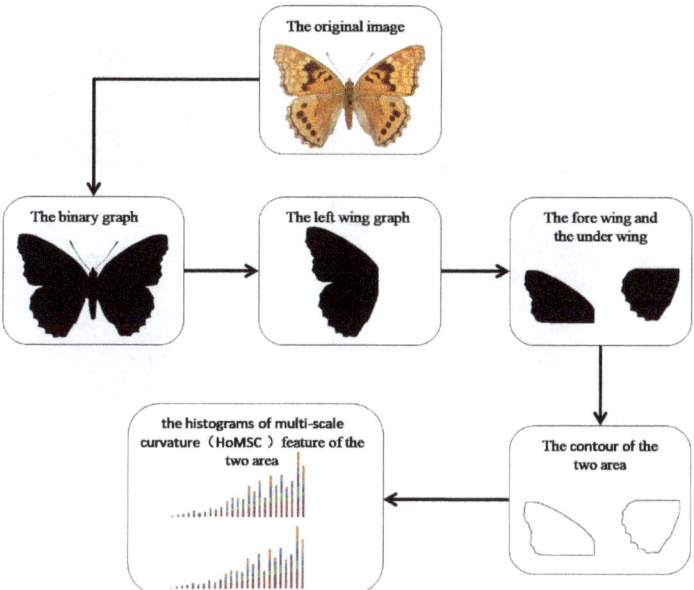

Fig. 2 Process of feature extraction

2.4 The Improved of HoMSC Feature

2.4.1 The Division and Reduction of Feature Extraction Area

The HoMSC method has been proved to be accurate and fast in describing complex contour features [8, 10]. However, the final result of this method is a set of statistics, which not contain any location information, and the location of the profile changes (forewings or hindwings) is often a key factor in determining different butterfly species. Therefore, according to the morphological structure characteristics of butterfly species, the feature extraction range is divided into forewings and hindwings (as Fig. 3 shows) for the final butterfly classification.

In addition, in our further study, we found that the significant changes are concentrated in only one section of the contour, so it is not necessary to use the HoMSC method to calculate the whole contour. Hence, the scope of the calculation can be narrowed down as Fig. 4 shows: Points $A(x, y)$ and $B(x', y')$ represent the highest and lowest points on a forewing or hindwing of the butterfly. From these two points, we can construct a circle that intersects the profile of the forewing or hindwing. The disjoint part of the circle is taken as the final contour curvature feature extraction area. Points $A(x, y)$ and $B(x', y')$ are the starting and ending points of the contour segment, respectively.

The left of a butterfly specimen

Forewing

Hindwings

Fig. 3 Division of feature extraction area

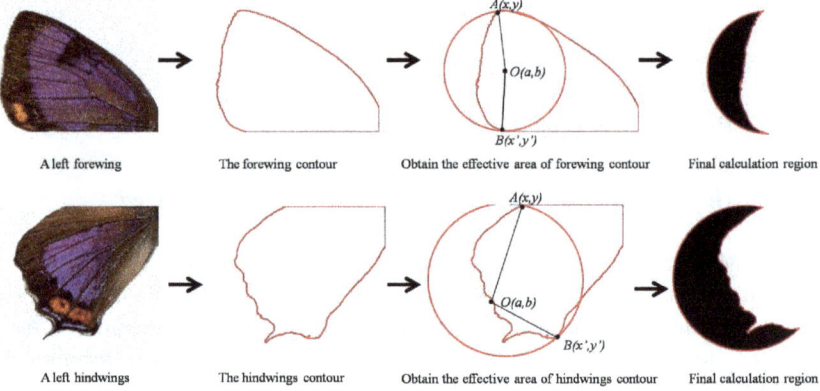

| A left forewing | The forewing contour | Obtain the effective area of forewing contour | Final calculation region |

| A left hindwings | The hindwings contour | Obtain the effective area of hindwings contour | Final calculation region |

Fig. 4 Reduction of feature extraction area

2.4.2 The Calculation of Contour Curvature

According to the method in [10], a contour curvature can be represented as an accumulated value $AC(P_{i(x,y)})$ in Eq. (1), while Eqs. (2) and (3) are satisfied:

$$AC(P_{i(x,y)}) = \sum_{j \in R} a_j(x', y'),$$ (1)

where

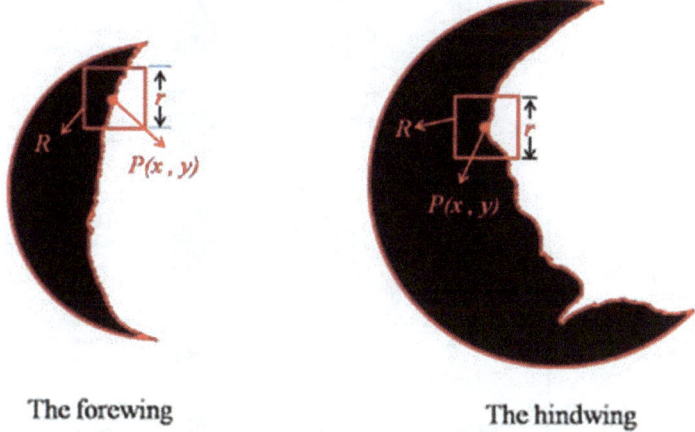

The forewing   The hindwing

Fig. 5 An example for calculating the curvature of front and rear wings

$$a_j\left(x', y'\right) = 1, \tag{2}$$

and

$$\sqrt{(x'-x)^2 - (y'-y)^2} < \frac{r}{2} \tag{3}$$

Here, point $P_{i(x,y)}$ is any point on the contour of a forewing or a hindwing. Taking the $P_{i(x,y)}$ as the center, searching for points within the radius $\frac{r}{2}$ of that satisfies the condition of Eq. (2) in the region R to accumulate. The values of $a_j\left(x', y'\right)$ is the pixel value of a point in the region R. As Fig. 5 show an example for calculating the curvature of front and rear wings.

According to this method, the intensity of the concave and convex change of the contour of a forewing (or a hindwing) can be reflected by the value of $AC\left(P_{i(x,y)}\right)$. If the value is less than $\frac{\pi r^2}{2}$, the contour at the point is concave, and the smaller the value, the more concave it is. Otherwise, the value represents the change of contour projection.

2.4.3   The Calculation of Multiscale

Although the above method provides us with a good help in calculating the curvature of the contour, if only a scale r is used to control the size of the calculated range r during the calculation, it cannot fully reflect the changes of the contour in details and the overall concave and convex. For this reason, we use a set of r values ranging from 3 to 28 pixels (a total of 25 scales) to achieve different computing scales. For each sampling point of the same contour, the curvature values at different scales are calculated first, and then all the calculated results are counted to form the final

Fig. 6 Shapes of three butterflies

HoMSC feature. As Fig. 6 shows, the three species of butterfly belong to the same family, which have some similarity in shapes. Among of them, Libytheana carinenta and Marpesia chiron differ greatly in the details of the outline, while Libytheana carinenta and Lycorea halia differ in the overall shape of the outline. Therefore, we can distinguish butterfly species with different details by a small scale and butterfly species with overall differences by a large scale.

3 Result

In this study, we discussed the effects of different types of features on butterfly classification problems and improve the histograms of multiscale curvature (HoMSC) calculation method to extract the shape features of wings (forewing and hindwing). To prove the effectiveness of this method, we use a weight-based k-nearest neighbor (KNN) search algorithm and 400 images of 20 butterfly species (which belong to six different families) for testing. The accuracy rate of this method reached 96%. The result suggested the improved HoMSC features can be efficient for the identification of butterfly species which are closely related.

For showing the performance of the new feature in this paper, we compared the original HoMSC features [9], the 13 geometrical features [11] and the branch length similarity entropy [12] are shown in Table 1. The four features were operated with the same classifier (k-nearest neighbor classifier) and image specimens.

Table 1 Performance of the improved HoMSC features, the HoMSC feature, the 13 geometrical features and the branch length similarity entropy

Features	Test image	Correctly identified images	Accuracy rate (%)
The improved HoMSC feature	200	192	96
The HoMSC feature	200	182	91
The 13 geometrical features	200	153	76.5
The branch length similarity entropy	200	159	79.5

4 Discussion

In this paper, the improvement of HoMSC feature extraction method is very effective. It can be seen from the experimental results that this improvement improves the accuracy of recognition, on the one hand, and reduces the amount of calculation, on the other hand. In addition, the results also indicate that the shape difference of butterfly wing can be used as an important feature to distinguish different butterfly species effectively in the classification tasks of different classification levels (family, genus, and species).

As a fine-grained image recognition problem, the classification of butterflies needs to be differentiated by the subtle differences among different species. Under the circumstance that the sample size is very limited, it is difficult for common deep learning methods to learn such subtle features. The method proposed in this paper can provide some help for solving such a specific task.

References

1. Pu, Z.Y., Zhou, D.Q., Yao, J., et al.: The living situation of biodiversity resource of China butterfly and a new conservation mode exploration. Ecol. Econ. **11**, 148–151 (2011)
2. Hopkins, G.W., Freckleton, R.P.: Declines in the numbers of amateur and professional taxonomists: implications for conservation. Anim. Conserv. **5**(3), 245–249 (2010)
3. Watson, A.T., O'neill, M.A., Kitching, I.J.: Automated identification of live moths (Macrolepidoptera) using digital automated identification system (DAISY). System. Biodivers. **1**(3), 287–300 (2004)
4. Wang, J.N., Lin, C.T., Ji, L.Q., et al.: A new automatic identification system of insect images at the order level. Knowl.-Based Syst. **33**(1), 102–110 (2012)
5. Kaya, Y., Kayci, L.: Application of artificial neural network for automatic detection of butterfly species using color and texture features. Vis. Comput. **30**(1), 71–79 (2014)
6. Hernández-serna, A., Jiménez-segura, L.F.: Automatic identification of species with neural networks. PeerJ **2**, e563 (2014)
7. Zhou, A.M., Ma, P.P., Xi, T.Y., et al.: Automatic identification of butterfly specimen images at the family level based on deep learning method. Acta Entomol. Sin. **60**(11), 1339–1348 (2017)
8. Li, F., Xiong, Y.: Automatic identification of butterfly species based on HoMSC and GLCMoIB. Vis. Comput. **33**(9), 1–9 (2017)

9. Xue, A.K., Li, F., Xiong, Y.: Automatic identification of butterfly species based on gray-level co-occurrence matrix features of image block. J. Shanghai Jiaotong Univ. (Sci.) **24**(2), 220–225 (2018)
10. Kumar, N., Belhumeur, P.N., Biswas, A., et al.: Leafsnap: a computer vision system for automatic plant species identification. In: 12th ECCV, pp. 502–516 (2012)
11. Zhang, J.W.: Automatic Identification of Butterflies Based on Computer Vision Technology. China Agricultural University, BeiJing (2006)
12. Kang, S.H., Jeon, W., Lee, S.H.: Butterfly species identification by branch length similarity entropy. Asia-Pac. Entomol. **15**, 437–441 (2012)

An Improved Neural Network Based on UNet for Surface Defect Segmentation

Daolei Wang and Yiteng Liu

Abstract Surface defect defection is a critical task in product quality assurance in industrial production lines. The defect characteristics of surface are easily affected by light. A surface defect detection model based on improved UNet-based neural network, surface defect UNet (S-UNet), is proposed to address the limitations of traditional detection algorithms in surface defect detection, as well as the problems of low accuracy, low precision, and cumbersome detection process. S-UNet uses atrous spatial pyramid pooling (ASPP) to acquire different receptive fields, overcoming the limitations of the UNet model with a single receptive field and improving the ability to segment targets of different sizes. S-UNet utilizes features at different levels to achieve complementary information and improve the coherence and accuracy of segmented areas. The model performs well on small industrial commutator datasets and steel surface defect dataset, reaching the optimal level in terms of precision, recall, and F-score. Compared to UNet and other models, as well as traditional methods, the proposed method achieves better results.

1 Introduction

Surface defects in industrial products have a negative impact on the aesthetics, comfort, and usability of the products, which is why manufacturers inspect the surface defects in order to detect and control them in a timely manner. Machine vision can largely overcome the drawbacks of manual inspection methods such as low sampling rate, low accuracy, poor real-time performance, low efficiency, and labor intensity and have been more and more widely studied and applied in modern industry.

In real and complex industrial environments, surface defect detection often faces many challenges, such as the existence of small differences between the defect image and background, low contrast, large-scale variations and types of defects, a lot of noise in the defect image, and even a lot of interference in the imaging of defects

D. Wang (✉) · Y. Liu
Shanghai University of Electric Power, Shanghai 200240, China
e-mail: alfredwdl@shiep.edu.cn

© The Author(s), under exclusive license to Springer Nature Singapore Pte Ltd. 2021   27
L. C. Jain et al. (eds.), *3D Imaging Technologies—Multidimensional Signal Processing and Deep Learning*, Smart Innovation, Systems and Technologies 236,
https://doi.org/10.1007/978-981-16-3180-1_4

in the natural environment, and so on. Landstrom et al. [1] used morphology to analyze billet surface images and extract defect features. Agarwal et al. [2] used the support vector machine method to identify hot rolled steel surface defects by extracting the morphology (shape, size, and location) of the image defects as features. Tsai et al. [3] used a two-dimensional Fourier transform to separate the normal and defective portions of a solar cell luminescence image for the purpose of extracting defective features. Zhang [4] constructs a five-scale, eight-directional Gabor filter set, and the original tile image is Gabor-transformed to obtain 40 subgraphs, and their mean and variance are used to characterize the tile. All the above methods require manual extraction of local or global features of defect images. However, local features are easily affected by noise, uneven illumination, complex background, and other factors, and global features cannot effectively detect the defects such as small cracks, rust spots, and small gaps, all of which increase the difficulty of manual extraction of features and is not conducive to handling complex classification problems and locating defective regions.

In recent years, with the successful application of deep learning models represented by convolutional neural networks (CNN) in many computer vision fields [5], such as face recognition, pedestrian re-recognition, scene text detection, target tracking, and automated driving, many deep learning-based defect detection methods have also been widely used in various industrial scenarios. Deep learning requires only data, does not require manual design of extracted features, and has the ability to adjust itself based on predictions.

Therefore, in this paper, we propose an improved UNet-based neural network S-UNet, which incorporates multiscale defect detection and segmentation network. S-UNet uses ASPP to acquire different receptive fields, overcoming the limitations of the UNet model with a single receptive field and improving the ability to segment targets of different sizes. S-UNet utilizes features at different levels to achieve complementary information and improve the coherence and accuracy of segmented areas.

2   S-UNet

Ronneberger et al. [6] proposed a lightweight UNet, which is a network mainly used for segmentation of biomedical images. In this paper, based on the UNet model, an improved design method for the S-UNet model is proposed, taking into account the characteristics of surface defect images of industrial products. S-UNet achieves a processing speed of 108 ms with good accuracy, and the embedding of ASPP enables the algorithm to be more robust to surface defects with varying scales.

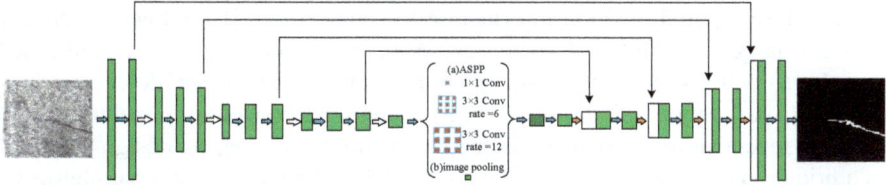

Fig. 1 Structure of the proposed S-UNet model

2.1 UNet

The core operation in UNet networks is to introduce hops between the encoding and decoding layers in order to reduce the loss of information of the underlying features caused by the pooling operation performed during the encoding phase. However, it is difficult to recover the spatial information contained in the low-level features even when the decoding phase performs the deconvolution operation on the high-level semantic features [7]. The high-level features help to classify pixels, while the low-level features help to generate fine boundaries. The hopping connection achieves high-quality segmentation by directly stitching together the low-level detailed features and high-level semantic features in the code-decoding stage and then compressing and fusing the features through the convolution operation.

UNet is arranged in a code–decode structure, with the encoder consisting of eight 3 × 3 convolutional cores and four maximum pooling layers and the decoder consisting of four convolutional and four deconvolutional layers, as shown in Fig. 1. In order to achieve nonlinearity and avoid overfitting effects, a ReLU activation function and a BN layer are added after each convolutional layer.

S-UNet decoder recovers highly abstracted high-level semantic features to the size of the input image. The S-UNet decoder is composed of four convolution kernels of 3 × 3 convolution layers, one convolution kernel of 1 × 1 convolution layers, and four deconvolution layers, in which the 3 × 3 convolution is used to obtain further fine segmentation results. The method in this paper is to establish four hop layer connections between the encoder and the decoder, to map the low-level features in the encoder directly to the decoder in a constant manner, and to fuse the spliced low-level features with the high-level features by 3 × 3 convolution. The final output of the S-UNet algorithm is a feature map with the same size as the original map and 64 channels downscaled to a single channel using a 1 × 1 convolutional layer, and the result is normalized to a probabilistic map using the SIGMOD.

2.2 ASPP

In UNet, downsampling the feature map using maximum pooling increases the receptive field of the convolution operation while keeping the convolution kernels small, but also loses spatial position information to some extent. If no downsampling is done and larger convolutional kernels are used, the parameter growth is multiplied. In order to better achieve the segmentation of multiscale surface defects, we add an atrous spatial pyramid pooling (ASPP) [8] and use convolution operations with different receptive fields to obtain rich contextual information, we design the ASPP as a bridge module between the encoding and decoding stages, and parallelize the expanding convolution with different expansion rates to extract more context. The features are then stitched together to take full advantage of the extracted multiscale features.

3 Experimental Evaluation

3.1 Datasets

Since most of the current deep learning-based surface defect detection methods are applied on specific datasets, many datasets are not publicly available, making it difficult to uniformly compare them. In order to better compare our algorithm with some classical algorithms on the same defect data, we have selected two very typical datasets as example datasets: The KolektorSDD dataset and steel surface defect dataset.

The KolektorSDD dataset is composed of images containing surface cracks on plastic electronic commutators. The dataset consists of 399 images collected in an industrial environment. Meanwhile, the steel surface defect dataset consists of 6666 images flat steel produced by industrial cameras containing surface defects.

3.2 Performance Measurement

In the open dataset experiments, in order to be able to analyze the test results from various perspectives and compare them with the models proposed in other experiments, we evaluate the prediction performance of the models using precision, recall, and F-score values.

$$\text{Precision} = \frac{TP}{TP + FP} \tag{1}$$

$$\text{Recall} = \frac{TP}{TP + FP} \tag{2}$$

$$F = \frac{2P \times R}{P + R} \tag{3}$$

3.3 Experimental Setup

In the study to identify surface defects, all experiments were conducted using Python 3.6, PyTorch as a deep learning library, and a platform of cuda 9.1 and cudnn 5.1 NVIDA Tesla K80 graphics processing unit (GPU) with 8 GB of memory on Google Cloud. The learning rate was initialized to 0.001, cross-entropy loss was used as a loss function, $\beta 1 = 0.9$, $\beta 2 = 0.999$, the same hyperparameters were used for each step, and the minibatch size was set to 8, using 64% of the images as the training set, 16% as the validation set, and the remaining 20% as the test set.

3.4 Data Augmentation

To illustrate the effectiveness of SCNet in surface defect detection, the model is compared with other models, including UNet, UNet + resnet50 [9], Deeplab v3 [8] + resnet50, and to illustrate the effectiveness of the method in this paper, the dataset is inspected by the traditional method of OTSU. Figure 2a–f shows some of the inspection results.

For the steel surface defect dataset, the Otsu effect is very poor, and the detection results contain a large number of false detections and background noise; when the defect shape in the image is more complex, there is a certain difference in the detection

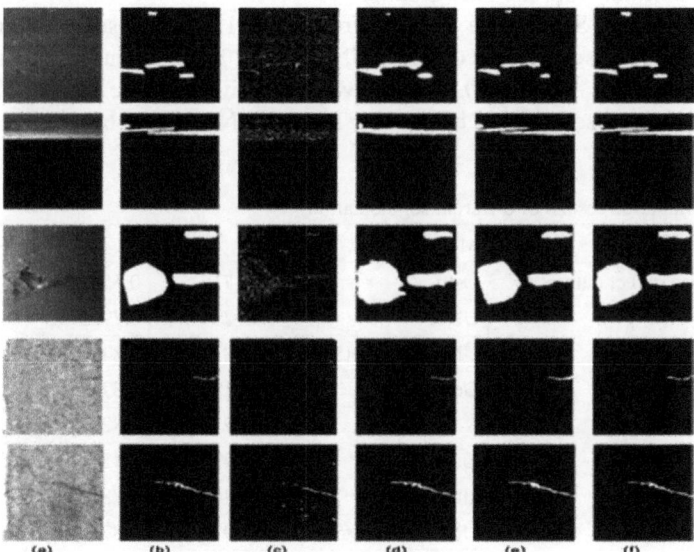

Fig. 2 First three rows of raw images are from the steel surface defect dataset. The last two rows of raw images are from the KolektorSDD dataset. **a** Image. **b** Image GT. **c** Ostu. **d** UNet. **e** Deeplab v3. **f** S-UNet

results of each model, as shown in Fig. 2, in the first and second rows of UNet, the defect segmentation area is sticky, the segmentation area is large, and there are cases of false detections and missed detections. Compared with other models, our proposed S-UNet has a clearer and more complete contour for defect segmentation, largely avoids misspecification of pixels, contains more feature information for detection, and has higher robustness and applicability.

For the KolektorSDD dataset, the Otsu method can roughly detect some defect shapes, but there are many over-detections. The three models based on deep learning can detect the defect shapes better when there is less interference in the image and the defects are clear. When the crack shape in the image is complex, as in the fifth line of Fig. 2, it is difficult for the human eye to distinguish the outline of the crack details, and UNet and Deeplab v3 misdetected the cracks, while the detected cracks were discontinuous. S-UNet, while showing some degree of misdetection, detects cracks more accurately and with more detail overall.

As shown in Table 1, the P and R of UNet are the lowest compared to the other three models, while we propose an improved UNet model with an ASPP structure that significantly improves the results by 26.8% and 4% for P and 11.7% and 9% for R, respectively, proving that the ASPP structure helps to improve the results of the deep learning model for defect detection. The accuracy of the traditional detection algorithm Otsu is not high for both datasets, and it performs better in KolektorSDD. The traditional detection algorithm is prone to segmenting a large amount of background noise as surface defects, which makes the detection results contain a large amount of false detection information and fails to effectively segment the defect contour from the background.

Both P and R of S-UNet are optimal. From Table 1, it can be seen that on the steel surface defect dataset, the F-score of S-UNet is 20.2%, 4.9%, and 1.1% higher than that of UNet, UNet + resnet50, Deeplab v3, respectively, on the KolektorSDD. The best results were also achieved by dataset. Thus, S-UNet has the best defect detection

Table 1 Evaluation results of different segmentation

Dataset	Method	P	R	F-score
Steel surface defect dataset	Ostu	0.1716	0.5557	0.2023
	UNet	0.7274	0.7933	0.7612
	UNet + resnet50	0.8835	0.8487	0.8717
	Deeplab V3	0.9185	0.8714	0.9043
	S-UNet	0.9224	0.8273	0.9150
KolektorSDD dataset	Ostu	0.5421	0.4277	0.4852
	UNet	0.8467	0.7802	0.8213
	UNet + resnet50	0.8926	0.8775	0.8850
	Deeplab V3	0.9322	0.9551	0.9427
	S-UNet	0.9457	0.9963	0.9703

results, while UNet has the worst crack detection results, which is consistent with the quantitative analysis.

4 Conclusion

Machine vision-based defect detection techniques are widely used in the industry, while traditional defect detection methods based on manual feature extraction rely heavily on the experience of engineers, making it difficult to identify the microscopic features of early defects. Convolutional neural networks (CNNs) have powerful feature extraction capabilities and can learn to adjust themselves based on the prediction results. In this paper, the proposed S-UNet converts defect images into pixel-level prediction masks (containing only defect pixels and background pixels) and quickly defect detection. Compared with traditional machine vision inspection methods UNet, DeepLabv3, and traditional machine vision inspection methods, the network inspection results proposed in this paper are more accurate and effective, which improves the performance of surface defect image inspection to a certain extent.

References

1. Landstrom, A., Thurley, M.J.: Morphology-based crack detection for steel slabs. IEEE J. Sel. Top. Sig. Process., 866–875 (2012)
2. Agarwal, K., Shivpuri, R., Zhu, Y., et al.: Process knowledge based multi-class support vector classification (PK-MSVM) approach for surface defects in hot rolling. Expert Syst. Appl. 7251–7262 (2011)
3. Tsai, D.M., Wu, S.C., Li, W.C.: Defect detection of solar cells in electroluminescence images using Fourier image reconstruction. Solar Energy Mater. Solar Cells, 250–262 (2012)
4. Zhang, Z.Y., Li, X., Bai, R.L.: The bias classification of magnetic tile surface defect based on the LSSVM. Opto-Electron. Eng., 78–83 (2013)
5. Weimer, D., Scholz-Reiter, B., Shpitalni, M.: Design of deep convolutional neural network architectures for automated feature extraction in industrial inspection. CIRP Ann., 417–420 (2016)
6. Ronneberger, O., Fischer, P., Brox, T.: U-net: Convolutional networks for biomedical image segmentation. In: International Conference on Medical Image Computing and Computer-Assisted Intervention, pp. 234–241. Springer, Cham (2015)
7. Drozdzal, M., Vorontsov, E., Chartrand, G., et al.: The importance of skip connections in biomedical image segmentation. In: Deep Learning and Data Labeling for Medical Applications. Springer, Cham, 179–187 (2016)
8. Chen, L.C., Papandreou, G., Schroff, F., et al.: Rethinking atrous convolution for semantic image segmentation. arXiv preprint arXiv:1706.05587 (2017)
9. He, K., Zhang, X., Ren, S., et al.: Deep residual learning for image recognition. In: Proceedings of the IEEE Conference on Computer Vision and Pattern Recognition, 770–778 (2016)

Conclusion

References

A Preliminary Study on the Segmentation Algorithm of Dongba Scripture

Yuting Yang and Houliang Kang

Abstract Dongba script is the only surviving hieroglyph in the world, and is known as the "living fossil" of writing. In 2003, the Dongba Scripture written in Dongba hieroglyph was included in the memory of the World Heritage List by UNESCO. Therefore, we realized an automatic segmentation algorithm for Dongba Scripture documents in sentence units. It improves the recursive polytomy algorithm based on projection. The algorithm is simple to implement, and is proved by experiments that it has high accuracy and robustness. And it provides more information for studying the layout rules, contextual connections, and the layout of Dongba Scriptures at a smaller level.

1 Introduction

Naxi hieroglyphs, called "Senjiulujiu" in Naxi language, can be literally translated as "prints left on wood and stone." [1] It is mainly used to write the Dongba Scripture that conveys the national culture by the Naxi master-Dongba, so people also call it Dongba hieroglyphs. Dongba hieroglyph is a very primitive pictograph, and it is also the only surviving hieroglyph in the world. It is known as the "living fossil" in the world of writing [2]. The Dongba Scripture written in Dongba hieroglyphs includes various aspects of Naxi social life, religious philosophy, historical evolution, folklore matters, science and technology, medicine, literature and art, and is an encyclopedia of the Naxi people. In 2003, Dongba Scripture was included in the Memory of the World Heritage List by UNESCO [3].

When Dongba Master wrote the Scripture, he first divided it into three or four lines with horizontal lines on the Dongba paper [1]. Then, he writes each line from top to bottom, from left to right, and uses vertical or double vertical lines to indicate

Y. Yang (✉)
School of Computer Engineering, Suzhou Vocational University, Suzhou 215000, Jiangsu, China
e-mail: tudou-yeah@163.com

H. Kang
Sports Department, Suzhou Vocational University, Suzhou 215000, Jiangsu, China

© The Author(s), under exclusive license to Springer Nature Singapore Pte Ltd. 2021 35
L. C. Jain et al. (eds.), *3D Imaging Technologies—Multidimensional Signal Processing and Deep Learning*, Smart Innovation, Systems and Technologies 236,
https://doi.org/10.1007/978-981-16-3180-1_5

Fig. 1 Excerpts from the
"Open road scripture "

the end of a sentence or paragraph [4]. Therefore, on the whole, Dongba hieroglyphs seem to be placed in an orderly table, and a cell of the table stores a sentence or a paragraph of Dongba Scripture, as shown in Fig. 1.

The writing characteristics of Dongba Scripture laid the foundation for us to use computer and image processing technology to extract the elements of the Scriptures in smaller units such as paragraphs and sentences. It also laid the foundation for us to establish different levels of Dongba Scripture database. Therefore, we combined the structural characteristics of Dongba Scripture and realized the automatic segmentation of Dongba Scripture documents in sentence units by improving the recursive polytomy algorithm based on projection. It can provide more information for studying the layout rules, contextual connections, and the layout of Dongba Scriptures at a smaller level.

2   An Automatic Segmentation Algorithm Based on Recursive Polytomy

Projection algorithm is an important algorithm widely used in document image layout analysis. Its basic idea is: for a document composed of pixels, if it is assumed that the background pixels of the image are white dots, the pixels corresponding to text, graphics, images, etc. are black dots. Then, the sum of the number of black dots in a pixel row (column) is its projection value, and the projection value of all pixel rows (columns) is the horizontal (vertical) projection of the entire document [5]. It is easy to find that the more text in a line and the denser the concentrated strokes and ink, the greater the projection value, while the projection value of the blank line is close to zero.

However, different from the tables in ordinary documents, Dongba Masters often used the lower border of the table cell as an auxiliary line when writing Dongba Scriptures. It integrates Dongba hieroglyphs with the lower border of the table, making the

Fig. 2 Dongba scriptures document and the projection value of each pixel row

dividing line of the table thicker and the projection value larger. And, the projection here shows a small range of peaks, as shown in Fig. 2. It can be seen that selecting the pixel row with a larger projection value as the dividing row can better realize the effective division of the Dongba Scriptures. Since each document contains at least 3–4 lines of Dongba Scriptures, if we choose the traditional recursive dichotomy algorithm, there may be incomplete segmentation. Therefore, we combine the layout characteristics of Dongba Scriptures and use a recursive polytomy algorithm based on projection. According to the projection value of each pixel row, it can find all the pixel rows that can be used for segmentation as much as possible, and realize the effective segmentation of all scripture lines.

2.1 The Recursive Polytomy Algorithm Based on Projection

The basic idea of the recursive polytomy algorithm is: Firstly, we binarize the document image and find the projection value of the horizontal pixel row in the document; Secondly, we find the widest trough (or peak) under a certain threshold, and divide the entire document into multiple sub-regions based on this boundary; Finally, we vertically project each sub-areas, and divide the scripture line by sentence as a unit.

However, if we only split the document based on the projection value of the pixel row, we may get some unexpected results. Take "Open Road Scriptures" as an example. Each page of the book includes 4 lines of Scriptures. If the upper and lower borders of the table are added, one page may include 6 dividing lines. If we directly choose 6 pixel-lines with the largest projection value as the dividing lines, the document may be over-segmented or incomplete. In addition, it may make more mistakes when judging the upper and lower borders of the scriptures, as shown in Fig. 3. The main reasons for these errors are: First, the irregular scanning of the scriptures, or the arbitrariness of the Dongba Master in writing the scriptures, makes it difficult to determine the number of dividing lines. Secondly, when Dongba Master wrote the Dongba Scripture, it was difficult to achieve the same height of each line; Finally, affected by the content of the scriptures or the esthetics of the layout, some

| (a) Over-segmentation of tables | (b) Invalid segmentation of table |

Fig. 3 Mis-segmentation of Dongba scripture. **a** Over-segmentation of tables. **b** Invalid segmentation of table

cells in the table have too many Dongba hieroglyphs. This causes the projection value of the pixel row to be too high and is wrongly judged as a dividing line.

2.2   The Improvement of Recursive Polytomy Algorithm

In order to achieve effective division of the Dongba Scripture, combined with the layout characteristics of the scriptures, we have added some restrictions during document division. Taking the "Open Road Scripture" as an example, we have added 3 restrictions.

1.   Based on the structural characteristics of the document, if we assume that the total height of each page is H and the number of lines is lineNum, then the height of the line is:

$$\text{height} = \frac{H}{\text{lineNum} + 1} \quad (\text{num} \geq 1) \tag{1}$$

   Among them, considering the fluctuation of row height, we take the denominator as lineNum + 1. In this way, the row height has a floatable interval, which enhances the robustness of the algorithm.

2.   In order to avoid the problem of over-segmentation or incomplete division of the document, we set the first and last lines as the upper and lower outer borders of the table, and the first and last columns are the left and right outer borders, as shown in Fig. 4. By adding borders around the table, we set the initial row and initial column that can be referenced for the division of the table. Therefore, when judging document segmentation lines, we still extract pixel lines in descending order of projection value. However, we use the upper outer line as the initial reference line, and only the pixel line with a height difference > =

Fig. 4 Add a border to the document table

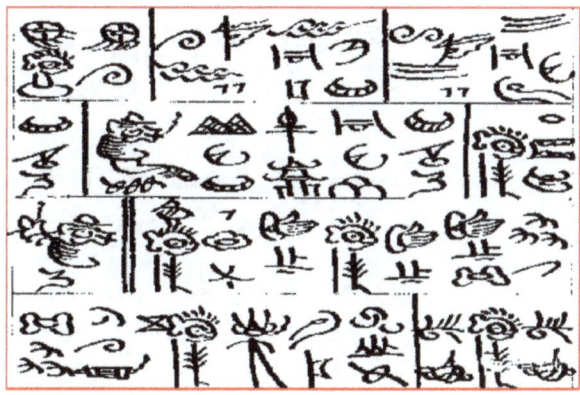

Fig. 5 Judgment and extraction of dividing lines

height and a larger projection value can be used as a dividing line, as shown in Fig. 5.

3. Compared with the dividing row, the projection value of the dividing column is also larger, but its position distribution is more random and has no regularity. However, its projection value is generally greater than 140. Therefore, we chose dividing column according to the projection value, and segment the Dongba Scripture into sentences or paragraphs, as shown in Fig. 6.

3 Test

3.1 Accuracy Test

In order to test the accuracy of the algorithm, we selected all 40-page documents in the "Open Road Scripture" to test the segmentation algorithm, and the results are

Fig. 6 Divide Dongba scripture into sentences or paragraphs

shown in Fig. 7. Among them, the average correct rate of row segmentation is 99.5%, the column segmentation is 95.79%, and the overall rate is 97.64%. It shows that our algorithm has high accuracy and good segmentation effect.

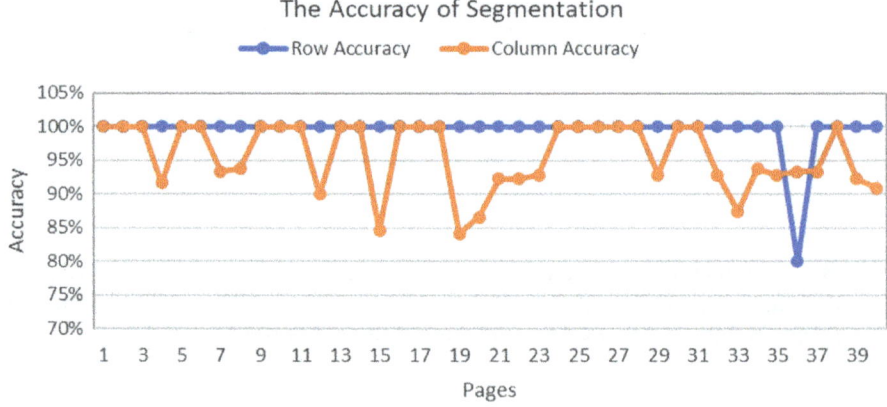

Fig. 7 The accuracy of segmentation

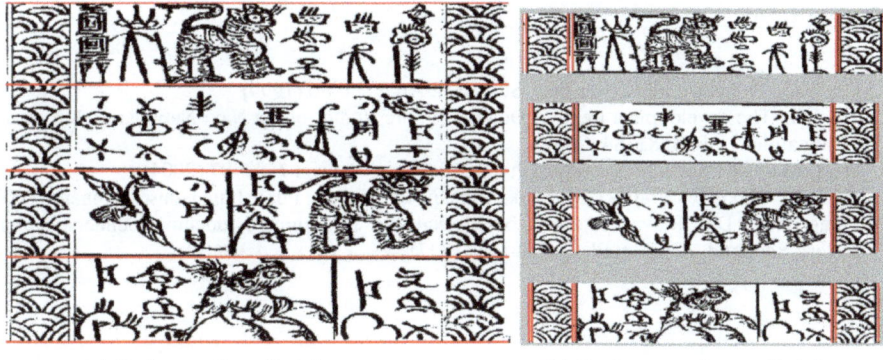

| (a) Scripture line division | (b) Scripture column division |

Fig. 8 Dongba Scripture segmentation with decorative borders. **a** Scripture line division. **b** Scripture column division

3.2 Robustness Test

In some Dongba Scriptures, some decorative marks are added to the first page to emphasize that the page is the starting page, as shown in Fig. 8a. In order to check whether the algorithm can overcome the influence of decoration marks, we choose this page for row and column segmentation test. As shown in Fig. 8a, b, when there are some decorations in the document, the algorithm can also accurately segment the rows and columns of the scripture, and mark the effective border of the decoration. It shows that our algorithm is robust.

4 Conclusion

We realized an automatic segmentation algorithm for Dongba Scripture documents in sentence units. It improves the recursive polytomy algorithm based on projection. The algorithm is simple to implement, and it is proved by experiments that it has high accuracy and robustness. And it laid the foundation for the study of Dongba Scriptures on a smaller level.

Acknowledgements This research was supported by the Talent Introduction Research Fund of Suzhou Vocational University (201905000034) and the Scientific Research Fund of Yunnan Education Department (2019J1152).

References

1. Ge, A.G.: Dongba culture overview. Stud. Nat. Art **2**, 71–80 (1999)
2. He, J.G.: The development trend of Dongba culture studies of the Naxi Nationality. J. Yunnan Nationalities Univ. **1**, 81–84 (2007)
3. He, L.M.: The transition of Dongba culture. Soc. Sci. Yunnan **1**, 83–87 (2004)
4. He, L.M.: Naxi Pictographs Copybook. Yunnan Nationalities Publishing House, Yunnan (2003)
5. Wang, H.Q., Dai, R.W.: Projection based recursive algorithm for document understanding. Pattern Recognit. Artif. Intell. **2**, 118–126 (1997)

Flower Gender Recognition Based on YOLO V4

Bao Zhou, Zhiwen Song, Yujin Wang, and Fangchao Hu

Abstract Flower gender recognition is essential for machine pollination in the field of intelligent agriculture. Then, the YOLO v4 algorithm is used to identify the gender of flowers, especially for pumpkin flowers. Though the adjustment of the model parameters and the number of classifiers, modification of prior anchors based on k-means algorithm, and also the introduction of transfer learning for freezing training, the gender recognition model acquired a mean average precision (mAP) of 94.58%, while the detection speed reached 43 frames/s. The experiment results showed that the model could effectively complete the flower gender recognition.

1 Introduction

Intelligent agriculture is a hot topic in the fields of agriculture and technology, as it can boost crop production, while saving labor. More and more advanced technologies are used in intelligent agriculture recently. Especially, many researchers use computer vision and deep learning to improve quality and efficiency of agriculture [1].

For example, Chung et al. [2] used computer vision technology to identify the diseased seedlings by quantifying the morphological and color characteristics of the seedling images captured from the flatbed scanner. Though the support vector machine (SVM) classifier, they acquire an accuracy of 87.9%. In order to improve the accuracy effectively, the CNN method has been used to modify the recognition algorithm, as it performs a preferable ability in speed, accuracy and robustness comparing with the traditional visual method. Zhang et al. [3] studied the crop disease identification method through the alteration of pooling combination, reduction in the number of classifiers, addition of the dropout operation and ReLU function. In this way, they modified the GoogleNet and Cifar10 models, and acquired accuracies of 98.9% and

B. Zhou · Z. Song · Y. Wang (✉) · F. Hu
School of Mechanical Engineering, Chongqing University of Technology, Chongqing 400054, China
e-mail: wangyujin@cqut.edu.cn

© The Author(s), under exclusive license to Springer Nature Singapore Pte Ltd. 2021
L. C. Jain et al. (eds.), *3D Imaging Technologies - Multidimensional Signal Processing and Deep Learning*, Smart Innovation, Systems and Technologies 236, https://doi.org/10.1007/978-981-16-3180-1_6

98.8%, respectively. Wu et al. [4] used channel pruned YOLO v4 deep learning algorithm to detect apple flower in natural environments. The results indicated a strong robustness to the changes of fruit tree varieties and illumination directions.

In addition, the flower gender recognition is also an important technology in intelligent agriculture, especially for the planting of cross-pollination crops, not only the monoecism but also dioecism. It could effectively shorten pollination time, avoid fruit dropping caused by pollination missing, and also reduce the cost of planting. Aiming to facilitate the pollination in the florescence by agricultural machine, we used the YOLO v4 algorithm to identify the flowers' gender, especially for pumpkin flowers.

2 Outline of YOLO V4 Algorithm

YOLO v4 is a regression-based object detection algorithm that consists of four major components, namely input, backbone, neck and head (Fig. 1). The object detector of YOLO v4 is composed of 161 layers of network, including convolutional layer, pooling layer and full connection layer.

Based on the above network structure of YOLO v4, the features of the input image are extracted by the feature extraction network, and, followed by grid partition operation. After the grid generation, the input image should be divided into S × S grids. Then, each grid is used to detect whether the objects are located in. If the number of detection categories is C, then each grid should be predicted by a number of anchor boxes. And the confidence of the anchor box can be calculated by Eq. (1).

$$
\begin{aligned}
\text{IoU} &= \frac{\text{area(box(Pred)} \cap \text{box(Truth))}}{\text{area(box(Pred)} \cup \text{box(Truth))}} \\
\text{Conf(Object)} &= \text{Pr(Object)} \times \text{IoU}
\end{aligned}
\tag{1}
$$

where, IoU represents the intersection-over-union between the prediction anchor box and the truth anchor box, while box(Pred) and box(Truth) represent the prediction anchor box and the truth anchor box, respectively. Conf(Object) represents the confidence of anchor box, Pr(Object) is used to indicate whether any object falls

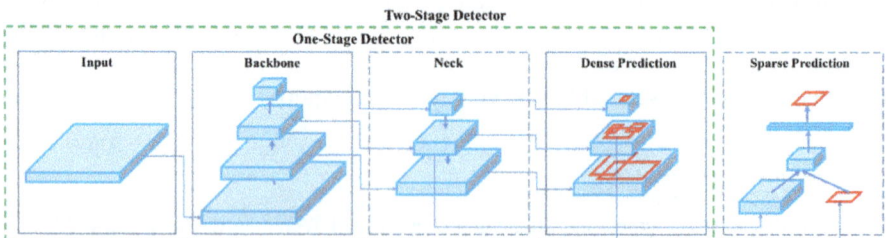

Fig. 1 Four major components of YOLO v4 [5]

Table 1 Configurations of the experimental platform

Hardware name	Type	Quantity
Mainboard	ASUS WSX299	1
CPU	Intel i9-10940X	1
Memory	KingstonDDR4 16 GB	4
GPU	GEFORCE GTX2080Ti	1
Disk	Kingston 500G	4

into the prediction anchor box, if any object falls into the prediction anchor box, Pr(Object) = 1, otherwise Pr(Object) = 0.

3 Our Works

3.1 Experimental Platform

The experimental operation was completed on a personal computer with the configurations shown in Table 1.

3.2 Experimental Operation

In order to identify the gender of the flower, the kind of pumpkin flower is used as an example due to the accessibility and universality in agriculture. By the web crawler technology, 3000 images of pumpkin flowers were collected from different search engines as data sets. Then, it is proportionately divided into three subsets, including training sets, verification sets and test sets, with the ratio of 8:1:1. The images in training sets and verification sets are labeled as "female" or "male" by manual work according to the gender.

Prior anchors are a series of anchor boxes with different width and height that play a foundational role in the target identification. The prior anchors of YOLO v4 are acquired according to COCO data sets that are not suitable for our experiment. In order to improve the identification precision, the sizes of prior anchors should be modified. Using the k-means clustering algorithm, we obtained 9 anchor boxes with the sizes gradually increase followed as $18 \times 23, 33 \times 43, 58 \times 70, 112 \times 127, 137 \times 202, 187 \times 252, 252 \times 310, 269 \times 191, 348 \times 349$.

After the acquirement of modified anchor boxes, model training is become to the fore. And the training parameters, such as batch size, epochs, IoU and learning rate (η), are very important for the training. Among them, the learning rate is constant in the traditional model training, resulting in more occupation of computing resource and time consumption. To speed up the model training and prevent the over-fitting

Table 2 Parameters setting of model training

Names	Batch size	η	Epochs	IoU
Values	8	0.001	1000	0.5
		0.0001	1000	

of the model in the case of insufficient data sets, the concept of transfer learning is introduced to set the learning rate. Namely, a learning rate is set for the previous half epochs, while another for the latter. The parameters of model training are shown in Table 2.

Cosine annealing algorithm is used during the learning rate scheduler that can be shown as Eq. (2).

$$\eta_t = \eta^i_{\min} + \frac{1}{2}\left(\eta^i_{\max} - \eta^i_{\min}\right)\left[1 + \cos\left(\frac{T_{\text{cur}}}{T_i}\pi\right)\right] \tag{2}$$

where, i represents the ith epoch, $\eta^i_{\max}$ and $\eta^i_{\min}$, respectively, represent the maximum and minimum values of learning rate. T_{cur} and T_i represent the current epoch and total epochs, respectively.

Based on the Cosine annealing algorithm, the learning rate is fine-tuned continuously. The loss of the model reduces quickly at the first, and then, tend to be convergent and stable after 200 epochs. The loss curve during training is shown as Fig. 2.

It can be seen, at the initial stage of the flower gender recognition model training, the model learning efficiency is higher, and the training curve convergence speed is faster. As the training deepened, the slope of the training curve gradually decrease. Finally, the model learning efficiency gradually reach saturation at the end of training, then the model output at the end of iterations can be used to identify the pumpkin flower gender.

Fig. 2 Diagram of loss value change

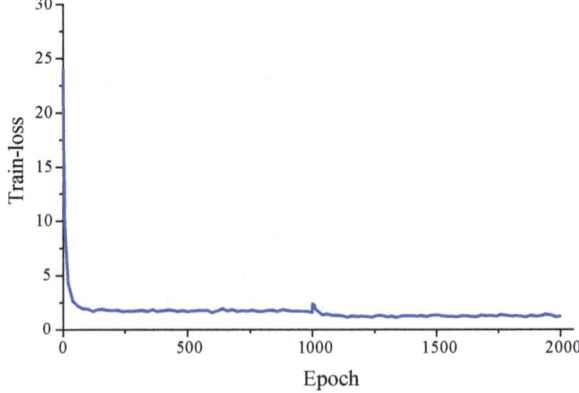

4 Results and Discussion

After the experimental operation mentioned above, flower gender recognition could be completed. The recognition results of pumpkin flowers are shown in Fig. 3.

The precision and recall are commonly used as indicators to assess the performances of the recognition model that can be calculated by Eq. (3).

$$P = \frac{TP}{TP+FP}$$
$$R = \frac{TP}{TP+FN}$$
(3)

where, P and R represent the precision and recall, respectively. TP, FP and FN represent the numbers of true positive cases, false positive cases and false negative cases, respectively.

Additionally, the mean average precision (mAP) [6] and detection speed are also used to reflect the global performance of the recognition model. The calculations of mAP and detection speed are shown in the following Eqs. (4)–(5).

$$\text{mAP} = \frac{1}{C} \sum_{i=1}^{C} \text{AP}(i)$$
(4)

$$\text{fps} = \frac{\text{NumFigure}}{\text{TotalTime}}$$
(5)

where, AP(i) is the average precision (AP) of ith category, C is the number of detection categories and $C = 2$ in this paper. The fps represents detection speed, NumFigure is the number of detection images, while TotalTime is the detection time.

According to Eqs. (3)–(5), the performances of the recognition model can be carried out. And the P-R curves are shown in Fig. 4.

The results reveal satisfactory performances of the flower gender recognition model, since that the mAP is 94.58%, while the detection speed can reach 43 frames/s. It means that the model not only ensure a high-detection speed but also high-detection accuracy.

Fig. 3 Flower gender recognition results. **a** Female flower, **b** male flower

(a): female flower　　　　　　　(b): male flower

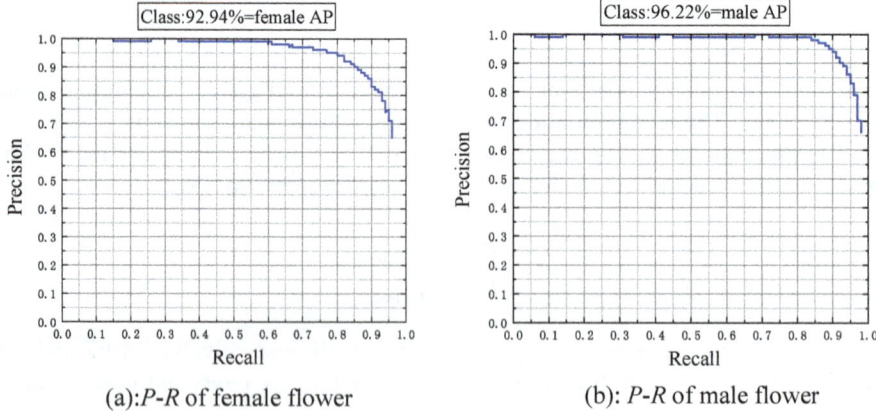

(a):*P-R* of female flower   (b): *P-R* of male flower

Fig. 4 The *P-R* curves of the flower gender recognition model. **a** *P-R* of female flower, **b** *P-R* of male flower

5 Summary

The gender of target flowers inputted from the camera can be quickly and accurately distinguished in our work through YOLOv4 algorithm. Experimental results show that the algorithm of flower gender recognition is effective and robust, although the kind of pumpkin flower is used as the sample merely. Therefore, our work can contribute to pollination by machine, especially in the intelligent agriculture field. However, the practical pollination situation in real scene is more complicated than it in the experiment, due to the problems caused by the interference factors, such as insufficient sunlight, partial occlusion, and also the shake of the target flower. The problems make the feature extraction quite difficult, which we should take effort to solve in future work.

Acknowledgements This work is supported by Scientific Research Foundation of Chongqing University of Technology (No. 2019ZD65), Student Research Project of Chongqing University of Technology (No. KLB20018), Project of Science and Technology Research Program of Banan District of Chongqing "Research on the key technology of single-track suspended self-propelled greenhouse operating robot based on machine vision".

References

1. Horng, G.J., Liu, M.X., Chen, C.C.: The smart image recognition mechanism for crop harvesting system in intelligent agriculture. IEEE Sens. J. **20**(5), 2766–2781 (2020)
2. Chung, C.L., Huang, K.J., Chen, S.Y.: Detecting Bakanae disease in rice seedlings by machine vision, Comput. Electron. Agric. **121**, 404–411 (2016)
3. Zhang, X.H., Yue, Q., Meng, F.F.: Identification of maize leaf diseases using improved deep convolutional neural networks. IEEE Access **6**, 30370–30377 (2018)

4. Wu, D.H., Lv, S.C., Jiang, M.: Using channel pruning-based YOLO v4 deep learning algorithm for the real-time and accurate detection of apple flowers in natural environments. Comput. Electron. Agric. **178**, 105742 (2020)
5. Bochkovskiy, A., Wang, C.Y., Liao, H.Y.M.: YOLOv4: optimal speed and accuracy of object detection, https://arxiv.org/abs/2004.10934. (2020). Last accessed 2020/4/23
6. Everingham, M., Gool, L.V., Williams, C.K.I.: The Pascal visual object classes (VOC) challenge. Int. J. Comput. Vis. **88**(2), 303–338, (2010)

Feature Understanding Based on Deconvolution Visualization

Weno Wu and Yun Pan

Abstract Convolutional neural network has been widely used in many fields, such as image recognition, target detection, image segmentation, and so on. But the process of training a neural network is very complex, which will consume a lot of time and resources. The process of input to output is similar to a "Black Box," and its specific working process cannot be seen. Visualization of the internal work of neural network is helpful to understand the working process of neural network. The method of deconvolution visualization is used to extract the features of the middle layer of the convolutional neural network and explain the extracted features, which can help to understand what features are extracted from each layer of the convolutional neural network. In this paper, the method of deconvolution visualization is further analyzed, Sobel operator and LBP operator are used to explain the extracted features, and a more detailed understanding of the internal network is obtained.

1 Introduction

As early as 1980, Fukushima Bangyan, a Japanese scientist, proposed a neural network structure including convolutional layer and pooling layer [1]. Further development was not achieved in the past 20 years until Yann Lecun proposed LeNet-5 in 1998 [2], which applies BP algorithm to the training of the neural network structure to form the prototype of contemporary convolutional neural network. In 2012, Alexanet mentioned by Hinton group introduced a new deep structure and dropout method in the ImageNet image recognition competition, which lowered the error rate from more than 25–15% [3]. So far, convolutional neural network has been widely used in various fields, and the feature map obtained from the network provides guidance for further research. For example, PhillipIsola et al. use the feature map to generate the confrontation network to realize image conversion [4], Tsung-Yi Lin et al. proposed

W. Wu · Y. Pan (✉)
School of Computer Science and Cybersecurity, Communication University of China, Beijing, China
e-mail: pany@cuc.edu.cn

© The Author(s), under exclusive license to Springer Nature Singapore Pte Ltd. 2021   51
L. C. Jain et al. (eds.), *3D Imaging Technologies—Multidimensional Signal Processing and Deep Learning*, Smart Innovation, Systems and Technologies 236,
https://doi.org/10.1007/978-981-16-3180-1_7

the method of feature fusion which is used for target detection to obtain more accurate semantic information [5]. Alexey Bochkovskiiy proposed YoLoV4 target detection network, in which feature graph is used for target detection and so on [6].

The process of training a neural network is very complex, which will consume a lot of time and resources. The process of input to output is similar to a "Black Box," and its specific working process cannot be seen. So, Matthew Zeiler proposed network visualization in 2013. Mattew et al. published a classic paper on neural network visualization [7]. The paper proposed that deconvolution can be used for visualization, and deconvolution visualization takes the feature map obtained from each layer as the input to deconvolve, and deconvolution results are obtained to display the feature map extracted from each layer. However, using deconvolution to visualize the network can only let us see the feature map intuitively, while it cannot explain what the extracted features are. Therefore, Sobel operator and LBP operator are used to examine the extracted features so as to better understand the training process of the network.

The paper introduces the basic principle of the method of deconvolution visualization in the beginning and selects a proper dataset to train the neural network. Then the paper shows part of the effect map and finally explains the extracted features.

2   Principle of Deconvolution Visualization

In order to understand how the neural network works and what features are learned in each layer of neural networks, Matthew D. Zeiler et al. used the method of deconvolution visualization to extract feature maps for visualization, that is, adding deconvolution process to the training of neural networks to achieve visualization.

Firstly, several concepts are introduced: unpooling, deactivation, deconvolution, and the flow chart of deconvolution visualization.

Unpooling: max-pooling is used in the pooling layer of the neural network, and the maximum activation value coordinates can be obtained after the max-pooling. In the process of unpooling, an all zero matrix with the same size as the pooling window is constructed. When the maximum activation value is located, the coordinate value is activated. This method is only an approximate process, after all, the positions of other coordinate values are not zero before (Fig. 1).

Deactivation: the purpose of the activation function using the ReLU function [as shown in (1)] is to ensure that the activation value of the output is positive, and the purpose of the reactivation function is to ensure that the activation value of the output is positive, so the function used is also a ReLU function.

$$\text{Re}LU(x) = x, x > 0, \text{Re}LU(x) = 0, x \leq 0 \tag{1}$$

Deconvolution (as shown in Fig. 2): also known as transposition convolution, deconvolution is actually the inverse process of convolution. In fact, deconvolution cannot restore the feature map to the previously input image. It can just restore the

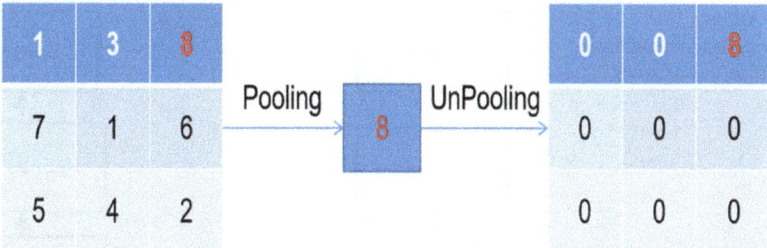

Fig. 1 Unpooling process

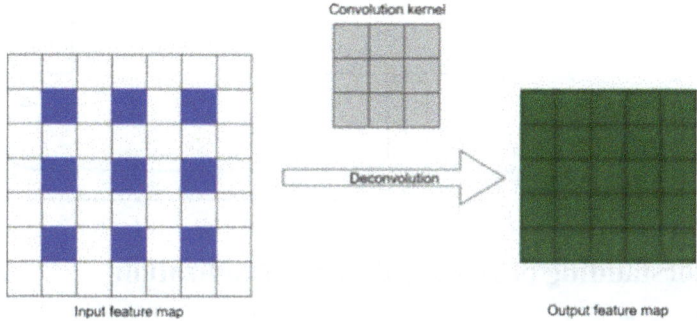

Fig. 2 Process of deconvolution

feature map to the size of the input image. The deconvolution results are obtained by deconvolution of the feature map, which is used to display the features extracted from each layer.

The size of the input feature map in Fig. 2 is 3×3 (input), the step size is $s = 1$, the pixel filling padding is 1, and 3×3 convolution kernel k is selected. Finally, the output size of deconvolution is calculated by Formula (2), and the result is 5×5.

$$\text{output} = s \times (\text{input} - 1) - 2 \times \text{padding} + k \tag{2}$$

In summary, a complete flow of deconvolution visualization can be shown in Fig. 3.

The advantage of deconvolution is that deconvolution visualization: deconvolution network can be used to visualize the process of convolution. Deconvolution of feature layer is helpful to understand how the continuous layer of neural network changes the input data. Through the visualization results, we can see the result (feature map) of input image after convolution. Each feature layer of convolution neural network contains N feature graphs (channel number), and each feature map corresponds to an independent feature. By visualizing all the feature maps, we can easily judge the performance of each convolution.

Fig. 3 Adding
deconvolution visualization
part into neural network

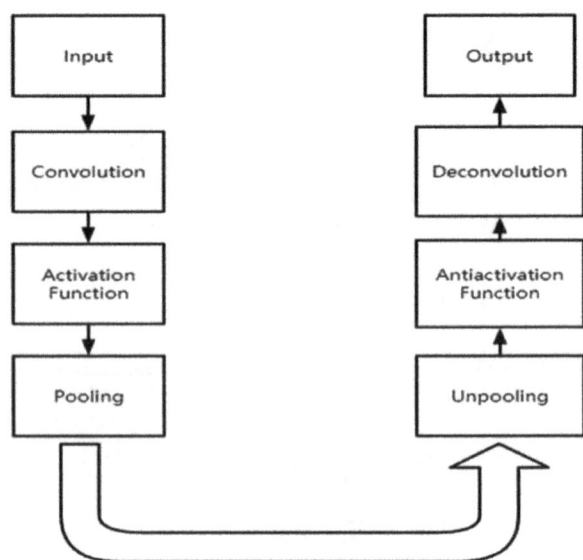

3 Understanding of Deconvolution Visualization

In order to understand the intermediate process of neural network, this section
will introduce deconvolution visualization to extract features. In this paper, VGG16
network is selected, and 3196 images are selected from www.goumin.com Web site.
Through deconvolution visualization, the features extracted from the network middle
layer can be more deeply understood.

3.1 Selection of Network Structure and Dataset

In order to reduce the consumption of resources and improve the accuracy of identification, this paper selects the VGG16 network structure (as shown in Fig. 4), because
the VGG16 network contains 16 convolution layers and full connection layers, which
is a relatively large network. However, its network structure is not very complex.
The whole network uses the same size of convolution kernel (3 × 3) and maximum
pooling size (2 × 2), and VGG16 uses a lot of 3 × 3 convolution concatenation.
The advantage is that convolution concatenation has less parameters than using a
large convolution kernel alone and has more nonlinear transformations than a single
convolution layer. For example, two 3 × 3 convolutions concatenation is equal to a 5
× 5 convolution. This paper uses the programming language Python 3.7, tensorflow
framework, and GTX1660Ti GPU.

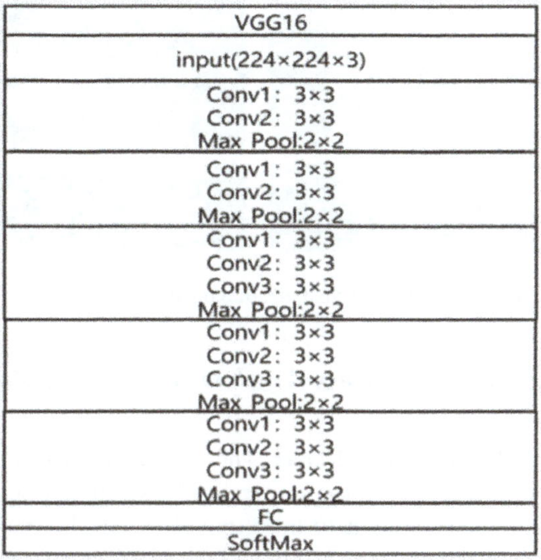

Fig. 4 VGG16 network
structure

VGG16
input(224×224×3)
Conv1: 3×3 Conv2: 3×3 Max Pool:2×2
Conv1: 3×3 Conv2: 3×3 Max Pool:2×2
Conv1: 3×3 Conv2: 3×3 Conv3: 3×3 Max Pool:2×2
Conv1: 3×3 Conv2: 3×3 Conv3: 3×3 Max Pool:2×2
Conv1: 3×3 Conv2: 3×3 Conv3: 3×3 Max Pool:2×2
FC
SoftMax

In order to show the visualization effect better, the dataset selected in this paper uses the dataset of a 224×224×3 picture of a dog collected from the Internet (as shown in Fig. 5).

Fig. 5 Picture of a dog

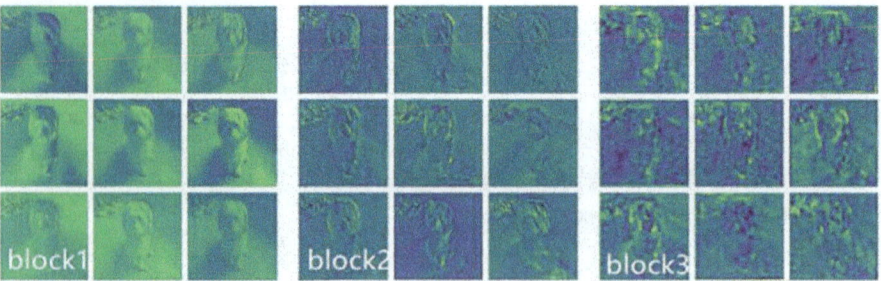

Fig. 6 Block1–block3 visualization results

3.2 *Effectiveness Analysis of Deconvolution Visualization*

Visualize the block 1 layer. The layer is the front convolution layer and extracts part of the feature map (as shown in Fig. 6). It can be found that the visualized feature map is very similar to the original image, because these feature maps outline the edge of the original image and the highlighted part of the feature map corresponds to the color of the whole block in the original image. Therefore, the features of this layer have the functions of edge detection and color detection.

Visualize the block 2 layer. Part of the feature map is extracted (as shown in Fig. 6), and the features in this layer are more complex than those in the previous layer. The features of this layer are similar to some texture, so this layer has the function of texture detection.

Visualize the block 3 layer. After extracting part of the feature map (as shown in Fig. 6), the feature of this layer is very complex, which is a high-level feature that cannot be expressed by language.

Finally, the block 5 layer is visualized. Part of the feature map is extracted (as shown in Fig. 7). After the image is convoluted for many times, the size of the image

Fig. 7 Block 5 layer visualization results

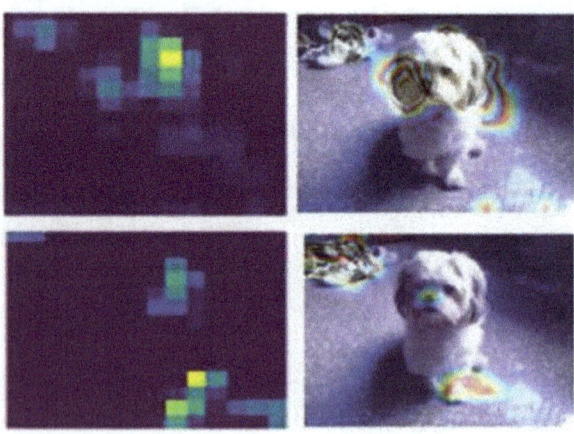

will become smaller and the image will be pixelated. After fusing these features with the original image, it is found that these features are some part of the dog. For example, the first picture shows that the head is activated, and the second image shows that the feet are activated. This layer is to extract more complex features, and then the network will make the final judgment according to these features.

It can be seen from the above figure that the activation value of the background in the feature map is very small. When the network learns the useful features, the activation value of the feature is very large. Therefore, the method of visualization method is to see what the network has learned. With the increasing number of network layers, the output image of each layer becomes more and more abstract, the features learned by the network become more and more advanced, and the extracted features will be less.

For the first two features, this paper uses the method of image processing to give further explanation. From the component contribution visualization [7], we can find that the final focus of the model is on the face. In this paper, Sobel operator edge detection method is used to extract the facial contour features of the original image (as shown in Fig. 8). The geometric moment Hu moment is used to calculate the similarity between the contour features and the features extracted from the first layer of the convolutional neural network to further illustrate that when the value of Hu moment is smaller, the two images or outlines are more similar.

Fig. 8 Extraction of contour features by Sobel operator

Fig. 9 Value of Hu moment

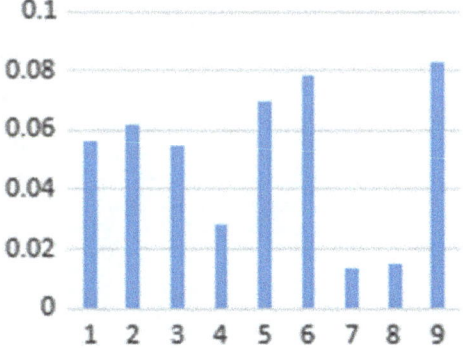

Fig. 10 Contour features
extracted by LBP operator

Figure 9 shows the Hu moments of the contour features extracted by Sobel operator
and the first layer features extracted by neural network. It can be seen that the features
extracted from the first layer of neural network are contour features.

In this paper, the improved circular LBP operator is used to extract texture features
(Fig. 10), which has rotation invariance. The biggest drawback of the basic LBP
operator is that it only covers a small area within a fixed radius, which obviously
cannot meet the needs of different sizes and frequency textures. In order to adapt to
the texture features of different scales and meet the requirements of grayscale and
rotation invariance, the 3 × 3 neighborhood is extended to any neighborhood, and
the square neighborhood is replaced by the circular neighborhood. The improved
LBP operator allows any number of pixels in the circular neighborhood with radius
R.

Figure 11 compares the similarity between the texture image extracted by LBP
operator and the feature map extracted from the second layer of neural network. It
can be concluded that the features extracted by the second layer of neural network
are texture features.

Fig. 11 Image similarity

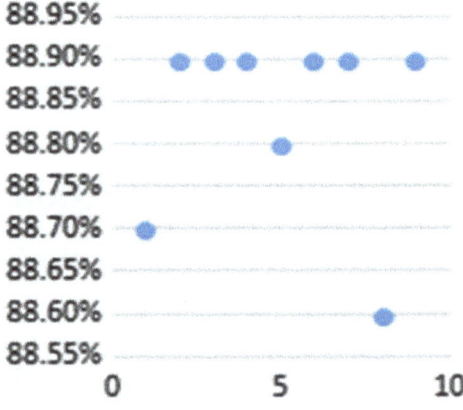

4   Conclusion

In this paper, deconvolution visualization is used to extract the features of convolutional neural network and Sobel operator and LBP operator are used to explain the extracted features. Through experiments, it is found that the first layer of the network focuses on the contour features of objects, the second layer focuses on texture features, and the features extracted from the third layer are unexplainable features, and the features extracted in the last layer of the network are few. It is found that the features extracted from the last layer of the network through visualization are the positional information of the object.

However, there are still some deficiencies that need to be further improved. For example, the features of the middle layer cannot be explained, and whether the extracted features can judge the quality of the network is also a difficulty at present. We hope that these difficulties can be solved in the future study.

References

1. Fukushima, K., Miyake, S.: Neocognitron: a self-organizing neural network model for a mechanism of visual pattern recognition. In: Competition and Cooperation in Neural Nets (pp. 267–285). Springer, Berlin, Heidelberg (1982)
2. LeCun, Y., Bottou, L., Bengio, Y., et al.: Gradient-based learning applied to document recognition. Proc. IEEE **86**(11), 2278–2324 (1998)
3. Krizhevsky, A., Sutskever, I., Hinton, G,E.: Imagenet classification with deep convolutional neural networks. In: Advances in Neural Information Processing Systems, 1097–1105 (2012)
4. Phillip, I., Jun-Yan, Z., Tinghui, Z., Alexei, A.E.: Image-to-Image Translation with Conditional Adversarial Networks. CVPR 2017, 5967–5976 (2017)
5. Tsung-Yi, L., Piotr, D., Ross, G., Kaiming, H., Bharath, H., Serge, B.: Feature Pyramid Networks for Object Detection. CVPR 2017, 939–944 (2017)
6. Alexey, B., Chien-Yao, W., Hong-Yuan, M.L.: YOLOv4: Optimal Speed and Accuracy of Object Detection.CoRR abs/2004, 10934 (2020)
7. Matthew, D., Rob, F.: Visualizing and understanding convolutional networks. ECCV **1**, 818–833 (2014)

Research on Resource Demand Prediction and Optimal Allocation Method in Cloud Computing Environment

Yixin Sun and Hailian Zhou

Abstract With the continuous development of digital informatization, many data processing technologies came into being. As a new data processing technology, the specific application environment of cloud computing is the whole Internet. Through data processing, analyze and sort out the data correlation, and provide relevant resources and services for users. Operating cloud computing infrastructure plays an important role in realizing the operation of cloud computing, so this paper mainly studies the companies that provide operating cloud computing and related companies that provide cloud computing services. Taking the optimization management of cloud computing resources as the research direction, the short-term dynamic prediction of cloud computing resources of load and other related means is used to conduct specific research. This paper intends to systematically describe cloud computing resources, and then effectively organize and rationally allocate cloud computing resources, improve resource service quality, reduce redundant costs of related enterprises, and achieve the best profit of enterprises.

1 Cloud Resource Load Feature Extraction and Classification

Cloud computing is a short-term resource load, which is usually presented in the form of time series, and these data are not linear but dynamic. Therefore, it is impossible to make accurate predictions and get ideal results only by using prediction models or prediction methods. At present, the mixed prediction mode is mainly used to deal with the data like time series with strong noise and dynamics. The main characteristics of this method are that it has a variety of models, detailed operation steps, and multiple effective methods. Its advantage lies in that it can accurately identify the potential system behavior patterns that are imperceptible among data. Through this model, we can quickly identify the corresponding method theory and classify the data with the

Y. Sun · H. Zhou (✉)
Weifang Engineering Vocational College, Weifang, China
e-mail: 294267909@qq.com

© The Author(s), under exclusive license to Springer Nature Singapore Pte Ltd. 2021
L. C. Jain et al. (eds.), *3D Imaging Technologies—Multidimensional Signal Processing and Deep Learning*, Smart Innovation, Systems and Technologies 236,
https://doi.org/10.1007/978-981-16-3180-1_8

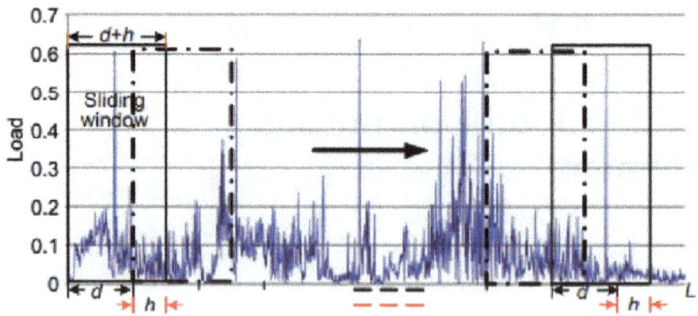

Fig. 1 Method of cloud computing resource load feature extraction based on FSOSW

same attributes. Therefore, based on this, an effective prediction can be achieved by selecting a suitable prediction model. [1]

An offline cloud resource load feature extraction method based on fixed-size overlapping sliding window (FSOSW) is shown in Fig. 1 [2].

It can be seen from Fig. 1 that l represents the time series of cloud computing resources. Using fixed-size overlapping sliding window technology, N subsequences with length d are intercepted on L, as shown by $n = l/d$. At the same time, n d-dimensional data points $xi(i = 1, 2, \ldots, n)$ are obtained, which can be used for clustering. That is, when clustering algorithm is used for feature classification, N subsequences with length d can be regarded as N D-dimensional data points [3].

According to the resource load subsequences with different lengths d and different load characteristics, the resource load subsequences are assigned to corresponding clusters by clustering method. In order not to affect the expected effect and the rationality of the return data, a cluster update mechanism with double time scales is adopted. Based on a short time scale (such as less than 6 h or 12 h), the similarity between new data and existing clusters can be calculated first. Then it belongs to the cluster with the greatest similarity. Based on a long time scale (such as more than 6 h or 12 h), the historical data and the new data are clustered together, and then the process is redefined. That is, the best cluster can be obtained by retraining the data. That is to say, by continuously changing and repeatedly training new data, the data obtained by this method can better adapt to the dynamic and real-time characteristics of cloud computing resource load [4].

2 Supervised Clustering Algorithm of Resource Load Characteristics Based on Kernel Fuzzy C-Means (KFCM)

Cluster analysis refers to grouping similar objects into multiple classes by dividing physical or abstract collections. Through this analysis, the objects with the same attributes can be grouped into the same class, and the contents in the class have common characteristics, while the objects outside the class have great differences. It can be understood that clustering is a method to study the potential relationship between objects from a logical or related perspective. Therefore, the analysis results can intuitively reflect the correlation and difference between related objects, which is an effective method to analyze and explore data [5].

Assuming that the cloud computing resource load subsequence set is represented by x, the n number sets can be expressed as $x = \{x_1, x_2, \ldots, x_n\}$, and $xi \in Rd$. It hao means that every data point in D-dimensional space is xi, where $I = 1, 2, \ldots, n$. If φ is used to represent the nonlinear kernel mapping function and H is used to represent the high-dimensional space, then the mapping of the input spatial data to the high-dimensional space is expressed as:

$$\Phi : Rd \rightarrow H, \ x \rightarrow \Phi(x) \tag{1}$$

At present, the commonly used kernel mapping functions φ are:
The polynomial kernel function:

$$K(xi, xj) = (xi, xj + c)d, c \geq 0, d \in N \tag{2}$$

The radial basis function (RBF) is:

$$K(x_i, x_j) = \exp\left(-\sum |x_i^a - x_j^a|^b / \sigma^2\right) \tag{3}$$

In which $a > 0$, $b > 0$, $\sigma > 0$, and $\sigma > 0$. When $a = 1$ and $b = 2$, RBF becomes Gaussian function.

Hypertangent kernel function:

$$K(x_i, x_j) = 1 - \tanh\left(-\|x_i - x_j\|^2 / \sigma^2\right) \tag{4}$$

where $\sigma > 0$.

In this paper, the radial basis kernel function is used as the mapping function in high-dimensional space. The main principle of KFCM clustering algorithm is to take the special spatial distance as the integration target and allocate the corresponding datasets to c clusters:

$$\min J_{KFCM} = \sum_{c=1}^{C} \sum_{i=1}^{N} u_{ic}^p \cdot \|\Phi(x_i) - \Phi_{v_c}\|^2 \tag{5}$$

By mapping function φ, the cluster center v_c of each cluster in feature space can be obtained. The corresponding calculation formula is as follows:

$$\Phi_{v_c} = \frac{\sum_{j=1}^{N} u_{jc}^{p} \Phi(x_j)}{\sum_{j=1}^{N} u_{jc}^{p}} \tag{6}$$

$\|\Phi_{(x_i)} - \Phi v_c\|^2$ represents the feature space distance, which can be calculated by the following formula:

$$
\begin{aligned}
&\left\| \Phi(x_i) - \Phi_{v_c} \right\|^2 \\
&= \left[\Phi(x_i) - \frac{\sum_{j=1}^{N} u_{jc}^{p} \Phi(x_j)}{\sum_{j=1}^{N} u_{jc}^{p}} \right] \times \left[\Phi(x_i) - \frac{\sum_{j=1}^{N} u_{jc}^{p} \Phi(x_j)}{\sum_{j=1}^{N} u_{jc}^{p}} \right]^T \\
&= \Phi(x_i) \cdot \Phi(x_i) - 2 \times \Phi(x_i) \times \\
&\quad \frac{\sum_{j=1}^{N} u_{jc}^{p} \Phi(x_j)}{\sum_{j=1}^{N} u_{jc}^{p}} + \frac{\sum_{j=1}^{N} u_{jc}^{p} \Phi(x_j)}{\sum_{j=1}^{N} u_{jc}^{p}} \times \frac{\sum_{j=1}^{N} u_{jc}^{p} \Phi(x_j)}{\sum_{j=1}^{N} u_{jc}^{p}} \\
&= K_{ii} - 2 \times \frac{\sum_{j=1}^{N} u_{jc}^{p} \Phi(x_j) \cdot K_{ij}}{\sum_{j=1}^{N} u_{jc}^{p}} + \frac{\sum_{m=1}^{M} \sum_{n=1}^{N} w_{mc}^{p} \cdot w_{nc}^{p} \cdot K_{mn}}{\sum_{m=1}^{M} \sum_{n=1}^{N} w_{mc}^{p} \cdot w_{nc}^{p}} \\
&= K(x_i, x_i) - 2K(x_i, x_j) + K(x_j, x_j) \tag{7}
\end{aligned}
$$

Finally, the optimal solution of the objective function can be obtained by using the expectation maximum (EM) method. That is, based on the constraint conditions, the optimal clustering center and the optimal membership matrix can be obtained, which are expressed by v_c and U, respectively. The specific formula is as follows:

$$v_c = \frac{\sum_{i=1}^{N} u_{ic}^{p} K(x_i, v_c) x_i}{\sum_{i=1}^{N} u_{ic}^{p} K(x_i, v_c)} \tag{8}$$

$$u_{ic} = \frac{(1/(1 - K(x_i, v_c)))^{1/(p-1)}}{\sum_{j=1}^{C} (1/(1 - K(x_i, v_c)))^{1/(p-1)}} \tag{9}$$

Through the above analysis, we can clearly understand the load characteristics of cloud computing resources and related feature classification [6]. The fixed-size overlapping sliding window technology based on the characteristics of cloud resource load classification is the main technical means. When dealing with the load characteristics of cloud resources, this paper mainly adopts the relevant extraction method and the kernel fuzzy C-means clustering method. Figure 2 is a multi-step forecasting framework for cloud resource load based on KFCM clustering.

Figure 2 shows the basic idea and specific operation steps of this paper, and the pseudocode of the specific algorithm, as shown in Table 1.

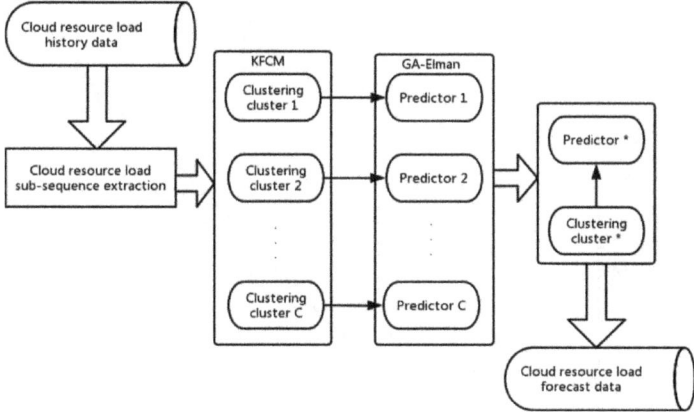

Fig. 2 Multi-step forecasting framework of cloud resource load based on KFCM clustering

Table 1 Multi-step forecasting algorithm of cloud resource load based on KFCM clustering

Algorithm KFCM-based GA-Elman Load Prediction Method in Cloud Computing (KGE)

Input: L cloud computing load time series data S_n, $n = 1, 2, \ldots, L$; Number of cluster C; Length of load subsequences need to be clustered d; Length of load needs to be predicted h; Current load subsequence x_N.

Output: Membership matrix U of each subsequence (d-dimensional data points) x_i ($i = 1, 2, \ldots, N$) in each cluster c ($c = 1, 2, \ldots, C$); Clustered data points in each cluster c; Predicted future cloud computing load PL.

1: Apply fixed size overlapping sliding window to extract the load subsequences from L historical data and construct N d-dimensional data points x_i for clustering, where $i = 1, 2, \ldots, N$.

2: For each data point x_i ($i = 1, 2 \ldots, N-1$) and each cluster c, calculate the membership matrix U by KFCM algorithm.

3: For each data x_i ($i = 1, 2 \ldots, N-1$), find one cluster in which it has maximum membership value from membership matrix U, and put this data into its maximum membership valued cluster.

4: For each cluster $c \in C$ do
    Feed each cluster a GA-Elman predictor;
    Put the data in cluster c and their next h-dimensional (length-h load time series data) data points *Ptrain* and *Ttrain* as training data into GA-Elman prediction model;
      Input current load subsequence x_N to the predictor.
    Train the GA-Elman prediction model;
  end

5: Find the best predictor from all clusters with minimum training error.

6: Obtain the optimal predicted future cloud computing CPU load PL from S_{L+1} to S_{L+h}.

3 Algorithm Verification

The historical data of Google cloud computing platform resources are used to verify the effectiveness of this method. Mean square error (MSE) is usually used to measure the validity of prediction performance.

Evaluate the performance stage of the model. The first step is to make clear the optimal number of clusters in all datasets. In this process, the predicted data point is 300 and the sliding window size is 600. That is, the prediction interval length

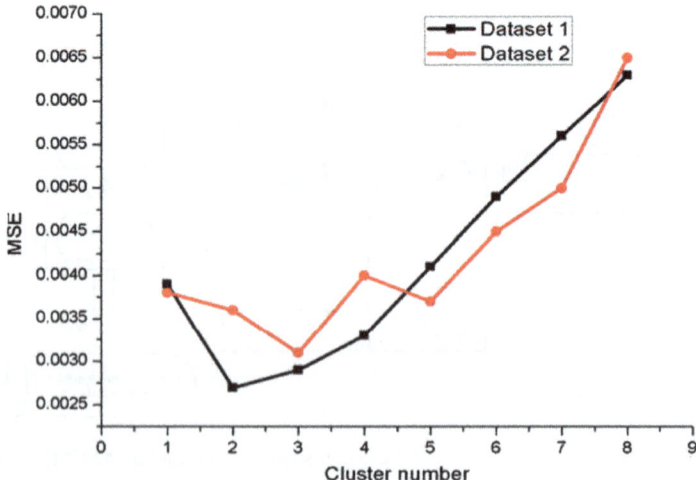

Fig. 3 MSE of prediction results of different cluster numbers

is half of the sliding window. Figure 3 gives MAE of GA-Elman prediction error corresponding to different cluster numbers from 1 to 8.

In the model built in this paper, the prediction error MAE values of different cluster numbers are all within the acceptable range, which shows that the model construction is effective. Mainly, when the number of clusters is 1, although the prediction system is not the best, it also meets the ideal prediction accuracy. Cluster analysis can classify resource load sets with the same attributes into the same cluster, which can not only find out the best parameters, but also help build the most accurate Elman neural network. In this study, the best number of clusters in dataset 1 and dataset 2 is 2 and 3, respectively. When the number of clusters increases, the prediction performance will decrease. It shows that the proper number of clusters can improve the prediction accuracy, while the excessive number of clusters will lead to the decline of prediction performance due to insufficient cluster training data.

4   Conclusion

In the cloud computing environment, resource management involves many problems, including the activities of resource representation, organization, discovery, matching, scheduling, configuration, and monitoring. These activities can have many positive influences and functions. For example, it can promote the stable operation of cloud computing system, make efficient use of resources, and improve service quality, so as to reduce the cost loss and increase the profits of enterprises related to cloud computing infrastructure and services. At the same time, it can optimize and integrate the relevant data resources of the data center reasonably, so as to improve the operational efficiency.

Based on the fixed-size sliding window technology, this paper identifies and obtains the load characteristics of cloud computing-related resources, and makes a prediction. In addition, it is also operated in a different way. That is to say, through its own dynamic characteristics, it extracts subsequences with adaptive and unequal length characteristics. In this way, the dynamic characteristics of cloud computing resource load can be obtained more intuitively and quickly. In addition, based on the above-technical means and related theoretical methods, we can also study non-equal length sequences. Usually, the algorithm used in this kind of sequence is unsupervised clustering algorithm. The model can also be selected and constructed according to the relevant characteristics of such data.

References

1. Zhou, P.: Summary of cloud computing related work of American government. Inf. Technol. Standard. **11**, 20–24 (2011)
2. EMC-Leading Cloud Computing, Big Data, and Trusted IT Solutions. http://www.emc.com, last accessed, 2019/11/21
3. Yu, F., Hua, W.N.: Bibliometric analysis of cloud computing research literature based on web of science database. New Century Libr. **7**, 57–63 (2013)
4. Foster, I., Zhao, Y., Raicu, I., Lu, S.: Cloud computing and grid computing 360-degree compared. In: 2008 Grid Computing Environments Workshop, pp. 1–10. IEEE (2008)
5. Shvachko, K., Kuang, H., Radia, S., Chansler, R.: The hadoop distributed file system. In: 2010 IEEE 26th Symposium on Mass Storage Systems and Technologies (MSST), pp. 1–10. IEEE (2010)
6. Chang, F., Dean, J., Ghemawat, S., Hsieh, W.C., Wallach, D.A., Burrows, M., Chandra, T., Fikes, A., Gruber, R.E.: Bigtable: a distributed storage system for structured data. ACM Trans. Comput. Syst. (TOCS) **26**(2), 1–26 (2008)

Research on Incentive Effect of Tax Preference on Enterprise Innovation Based on DEA Model

Yunjing He

Abstract The innovation of enterprises can promote the rapid development of economy and adjust the economic structure. In order to establish sustainable competitiveness, it is essential for enterprises to innovate constantly in the process of development. This paper will study the incentive effect of tax incentives on enterprise innovation based on DEA model (data envelopment analysis method), starting with the theoretical content of incentive policy and enterprise innovation performance, and then complete the subsequent DEA model establishment based on this, and finally, put forward relevant strategies to enhance the incentive policy of innovation tax in China.

1 Theoretical Research on Tax Incentive Policy

1.1 Theoretical Research on Incentive Policies

The research of policy incentive theory starts with its effect evaluation and focuses on two aspects of economic growth and market failure when using METC (marginal efficiency) method. The above two aspects are relatively mature in theory. In order to better measure the intensity of tax policy, the three main methods used at present are to measure whether the tax can achieve the optimal level of benefit, to complete the loss of benefit and tax collection by increasing the balance, and to compare the expenditure and subsidy. All the above three methods can complete the government's measurement of the intensity of corporate tax incentives and compare them with the international average level. Among them, the marginal efficiency METC method is mostly used by Chinese enterprises at present, and its specific calculation formula is:

This paper shows the details of different tax credit policies and summarizes them in the form of charts. See Table 1 for details.

Y. He (✉)
Business School of Henan University, Henan, China

© The Author(s), under exclusive license to Springer Nature Singapore Pte Ltd. 2021
L. C. Jain et al. (eds.), *3D Imaging Technologies—Multidimensional Signal Processing and Deep Learning*, Smart Innovation, Systems and Technologies 236,
https://doi.org/10.1007/978-981-16-3180-1_9

Table 1 METC details of tax credit policy

	Total tax credit	Incremental tax credits	
		Rolling base	Fixed base
METC	All expenditures are deductible at the rate of tax exemption	When the expenditure increment is less than or equal to 0, METC is less than 0 and produces negative effect	The expenditure below the fixed base is METC equals 0

1.2 Theoretical Research on Enterprise Innovation Performance

There are two kinds of factors that will influence the innovation performance of enterprises, external factors, and internal factors. Among the internal factors, the first is the network structure of enterprises and the construction of external social networks. Secondly, technological innovation and technological system reform are also very important. With the development of technology, technological opportunities are increasing, and the possibility of technological leap and progress will also increase. And the government's behavior decision making is also classified into external factors. In terms of internal factors, firstly, the enterprise's human resources, secondly, the enterprise's external knowledge absorption ability and knowledge integration mechanism are also very important influencing factors. Through the research, it can be found that when the rated absorptive capacity of enterprises is strong, the methods of knowledge transfer and knowledge integration are used to complete technological innovation.

2 DEA Model of the Influence of Tax Incentive Policy on Enterprise Innovation Performance

2.1 Tax Incentives Affect Innovation Performance DEA Model

From the above, it can be seen that enterprises will have positive spillover effect in the process of innovation activities, which leads to a large number of enterprises unable to reap all the benefits of innovation investment. Therefore, in the process of innovation activities, the social benefits of enterprises will be greater than the private benefits. Therefore, this paper calculates the DEA model, and the details of its change chart are shown in Fig. 1 [1].

After the reduction of influencing factors, it can be found that enterprises will increase the use of capital when the capital is low, and the calculation formula of DEA model diagram in this case is:

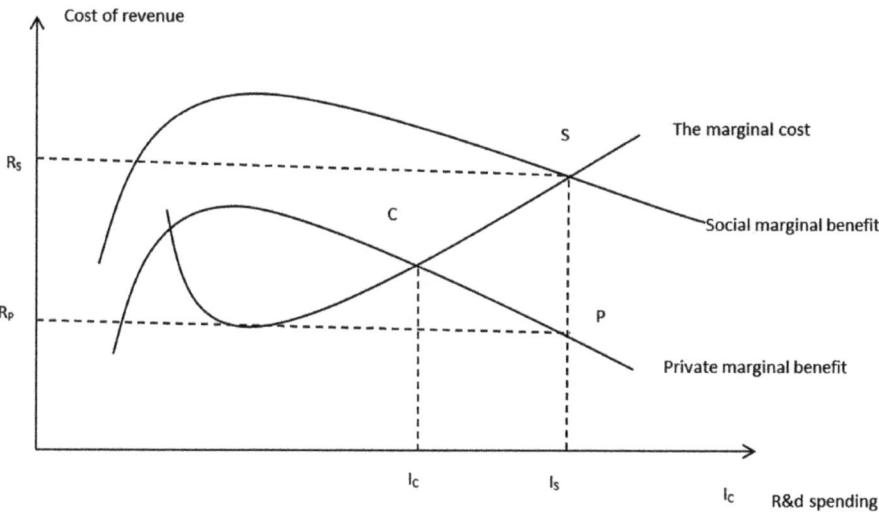

Fig. 1 Income diagram of DEA model of enterprise innovation activities

$$C = q(r + d)(1 - uz - uy - uh - ug)/(1 - u)$$

In the above formula, C is used as the cost of capital, Q is used to maintain the price of capital goods, and R is the market interest rate; D is the depreciation rate. Uz are investment funds, uy pre-tax assets, uh capital R & D fundamentals, ug after-tax assets. It can be found that the depreciation rate of investment funds and costs and the intuitive comparison of assets show an inverse trend, that is, the larger the amount of items deducted before tax, the smaller the capital cost, which can sum up the changes of DEA model of the impact of tax incentive policies on enterprise R&D funds. For details, see Fig. 2.

2.2 Research Hypothesis

The effect of innovation tax incentive policy on enterprise innovation performance is studied by endogenous growth theory. It can be known that economy cannot only rely on external forces to promote growth, even if its short-term growth trend appears, it is not sustainable. In the follow-up study, it can be found that the country needs to educate and cultivate the national innovation framework system when China's policies are integrated into it. At the same time, it can be found that the government still occupies a very important position in the process of promoting innovation, and in the process of innovation activities and even the subsequent market-oriented sales, it can directly participate in the sales process of products, which not only provides technical support, but also improves the subsequent environmental support [2].

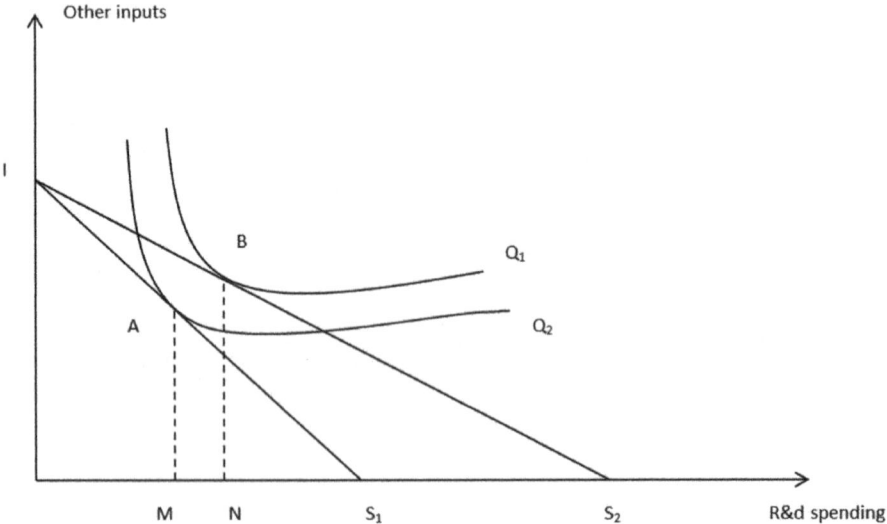

Fig. 2 DEA model diagram of the influence of tax incentive policy on R&D capital investment of enterprises

3 Strategies of Promoting Innovation Tax Incentive Policy in China

3.1 Policy Utilization

From the above content and the calculation of DEA model function, it can be found that the formulation and implementation of tax incentive policies need a long time to summarize, so enterprises also need to adapt to the process. Therefore, the use of policies should be considered from the following three points. First, it is necessary to invest in enterprises supported by the state. From the above, it can be found that most developed countries currently use universal preferential policies and have not set up corresponding industrial settings, so China can also develop in the direction of high-tech, environmental protection, and comprehensive utilization of resources. Secondly, although our country has begun to develop toward weakening the regional preferential tax, there will inevitably be a tax tilt in the process of adaptation. Therefore, enterprises in special circumstances can consider using tax savings to provide a better development environment [3, 4].

3.2 Policy Implementation

The difficulty in the operation of tax preferential policies may be the key factor affecting them. The procedures for enterprises to apply for tax preferential policies are usually complicated, and there may be difficulties in the implementation process, which will greatly reduce the enthusiasm of enterprises to apply. For example, high-tech enterprises will involve the financial department, R&D department, and human resources department in the application process. At this time, in the process of financial summary and report, it is not only necessary to perform independent evaluation, but also the preferential policies of enterprises are relatively complex on the other hand. Therefore, the government should simplify the application form and process, constantly increase the intensity of publicity, compile the government's policies that are conducive to enterprises' application for innovation investment into a book, and publicize them on major publicity platforms [5–9].

4 Conclusion

Through the discussion in this article, it can be proved that tax incentive policies have obvious innovation incentive effects on Chinese enterprises. In the specific conclusions, it can be found that tax incentive policies are directly proportional to the input cost of enterprise innovation activities. By calculating the content of the DEA model, we found that with the promotion of preferential tax policies, the incentive effect will be more obvious. Therefore, it is necessary to emphasize the enterprise's cognition of innovation behavior, and continuously collect panel data through continuous comparison and summary, in order to further improve the dynamic relationship between tax preferential policies and enterprise innovation performance.

References

1. Dietl, H.M., Grossmann, M., Lang, M., Wey, S.: Incentive effects of bonus taxes in a principal-agent model. J. Econ. Behav. Organ. **89**(2), 93–104 (2012)
2. Maynard, B.R., Kjellstrand, E.K., Thompson, A.M.: Effects of check and connect on attendance, behavior, and academics: a randomized effectiveness trial. Res. Soc. Work. Pract. **24**(3), 296–309 (2014)
3. Erdfelder, E., Castela, M., Michalkiewicz, M., Heck, D.W.: The advantages of model fitting compared to model simulation in research on preference construction. Front. Psychol. **6**(140), 140 (2015)
4. Lee, K., Go, D., Park, I., Yoon, B.: Exploring suitable technology for small and medium-sized enterprises (SMEs) based on a hidden Markov model using patent information and value chain analysis. Sustainability **9**(7), 1100 (2017)
5. Mikkola, H., Häkkinen, U.: The effects of case-based pricing on length of stay for common surgical procedures. J. Health Serv. Res. Policy **7**(2), 90–97 (2002)

6. Cappelen, A., Raknerud, A., Rybalka, M:. The effects of R&D tax credits on patenting and innovations. Res. Policy **41**(2), 334–345 (2012)
7. Parker, S.L., Jimmieson, N.L., Techakesari, P.: Using stress and resource theories to examine the incentive effects of a performance-based extrinsic reward. Hum. Perform. **30**(4), 169–192 (2017)
8. Molenaers, N., Nijs, L.: Why the European Commission fails to adhere to the principles of good donorship: the case of the governance incentive tranche. Eur. J. Dev. Res. **23**(3), 409–425 (2011)
9. Sinai, P.T.: Part 2: economic analysis of tax expenditures. Revenue costs and incentive effects of the mortgage interest deduction for owner-occupied housing. Nat. Tax J. **64**(2), 531–564 (2011)

The Development Path of China's Social Security Helping Rural Poverty Alleviation Based on Multi-dimensional Integration Model

Chen Wang

Abstract With the development of China's social security system, China's social security system has gradually improved, playing an important role in ensuring basic needs, narrowing the income gap and reducing social risks. Although the coverage of social system is expanding, there are still some problems, such as low-security standard, obvious gap between urban and rural areas and regions, fragmentation of policy, lack of foresight and initiative. These problems restrict the realization of anti-poverty effect of rural social security system to a great extent. In view of this related problem, this paper collects 11 contiguous poverty-stricken areas and 3 areas with special poverty alleviation policies. Based on the data of 14 poverty-stricken areas, the weight of multi-dimensional poverty identification index system is studied through the establishment of multi-dimensional poverty identification model and the selected poverty-stricken counties to further analyze the specific situation and further improve the social security system.

1 The Seriousness of Rural Poverty

From income poverty to ability poverty to multi-dimensional poverty measurement methods, people from all walks of life have deepened their understanding of poverty, and the strategies to deal with poverty are constantly enriched and improved. Under the guidance of precise poverty alleviation strategy, China's rural poverty alleviation has achieved remarkable results. The number of rural poor people in China decreased year by year from 2012 to 2019.

C. Wang (✉)
Southwestern University of Finance and Economics, Sichuan, China
e-mail: yanpingzhang2020@163.com

© The Author(s), under exclusive license to Springer Nature Singapore Pte Ltd. 2021
L. C. Jain et al. (eds.), *3D Imaging Technologies—Multidimensional Signal Processing and Deep Learning*, Smart Innovation, Systems and Technologies 236,
https://doi.org/10.1007/978-981-16-3180-1_10

2 Distribution of Poverty-Stricken Areas

2.1 Distribution of Concentrated Contiguous Poverty-Stricken Areas

In 2012, the State Council formulated the Outline of China's Rural Poverty Alleviation and Development (2012–2019) (hereinafter referred to as the Outline) aiming at
accelerating the development of poverty-stricken areas and achieving common pros-
perity. The outline delineates 11 contiguous destitute areas and 3 areas with special
poverty alleviation policies. Refer to Table 1 for the related data of 14 special hard-
ship areas. It can be seen from Table 1 that the average annual precipitation in Tibetan
areas of the four provinces is the largest, and there is a huge difference with other
extremely poor areas. As a result, natural disasters related to water, such as floods
and mudslides, frequently occur in the area, which makes the material property in the
area unable to accumulate for a long time, thus reducing the sustainable development
capacity of the population. The object of this study is to select the relevant areas in
Tibetan areas of four provinces, aiming at improving the survival and development
ability of the population in the study area from its own perspective, and providing
suggestions for overcoming poverty and getting rid of poverty compared with this
area [1–3].

2.2 Multi-dimensional Poverty Identification Model

In the process of multi-dimensional poverty measurement in the concentrated areas
of poor villages in Xiaojin County, the multi-dimensional poverty index reflects
the poverty dimensions of the administrative villages in the study area and the
poverty depth of each dimension, revealing the "orderly" or "disordered" state in
the system development process of concentrated areas of poor villages, and making
future poverty alleviation strategies by using this state analysis. Multi-dimensional
poverty measurement model in the concentrated areas of poor villages in Xiaojin
County will be derived by the comprehensive index method of multiple indicators
[4–7].

(1) the dimensionless index

$X = [Y_{ij}]$ is a $n * d$ dimension matrix, and the element Y_{ij} represents the value
of the i in the individual dimension. Because the indexes selected in the study are
of different dimensions, dimensionless processing is required before substituting the
data into the operation, and the processing formula is as follows:

$$Y_{ij} = \frac{X_{ij} - X_{\min}}{X_{\max} - X_{\min}} \tag{1}$$

Table 1 Distribution of concentrated contiguous poverty-stricken areas

The national concentrated contiguous zone of extreme poverty	Gross area (10^4 km^2)	Average annual precipitation (ml)	Total population (10,000)	People in poverty (10,000)	Incidence of rural poverty (%)	Number of counties (per unit)
Liupanshan area	16.6	422	1817.7	349	19.2	69
Qinba mountainous area	22.5	875	2707.3	444	16.4	80
Wuling mountainous area	17.18	1700	2810.7	475	16.9	71
WuMengShan district	10.7	384	2055.8	442	21.5	38
Rocky desertification region of Guangxi and Guizhou	22.8	1436	2637.8	488	18.5	91
Western Yunnan border mountains	20.9	1100	1256.5	240	19.1	61
The southern foothills of the Greater Hinggan Mountains	14.5	404	528.6	74	14	22
Yanshan—Taihang mountain area	9.3	440	892.9	150	16.8	33
Luliang mountains	3.6	462	343.6	67	19.5	20
Dabie Mountain area	6.7	1069	3266.7	392	12	36
Luoxiao mountains	5.3	1640	937.1	134	14.3	24
Tibet area	122.8	488	257.4	61	23.7	74
Four provinces Tibetan areas	109.5	7018	425.6	103	24.2	77
Xinjiang Southern Xinjiang three prefectures	45.9	1766	526.6	99	18.8	24

$$Y_{ij} = \frac{X_{\max} - X_{ij}}{X_{\max} - X_{\min}} \tag{2}$$

(2) Weight calculation

In the study, the weights of indicators in different dimensions in poor villages are relative, which reflects the importance of an indicator to the research object compared with other indicators. P_{ij} is the weight of dimension j and the critical value of poverty.

① Calculate the proportion of the ith sample village to the $\sum_{i=1}^{n} y_{ij}$ index under the j index

$$P_{ij} = \frac{Y_{ij}}{\sum_{i=1}^{n} Y_{ij}} \tag{3}$$

② Calculate the entropy value of the j index, $P_{ij} \ln P_{ij}$ series vector. e for the poor

$$e_j = -\frac{1}{\ln n} \sum_{i=1}^{n} P_{ij} \ln(P_{ij}), \quad (j = 1, 2, \ldots m) \tag{4}$$

③ Calculate the entropy weight ω_j of the index

$$\omega_j = \frac{1 - e_j}{m - \sum_{j=1}^{n} e_j} \tag{5}$$

In this study, entropy method is used to give weight to multi-dimensional poverty identification indicators in the concentrated areas of poor villages in Xiaojin County. Entropy method mainly distributes the weight of indicators' overall impact on the system. Its basic principle is that the greater the entropy, the less information, and the smaller the utility value and weight of indicators. Calculate the weight of multi-dimensional poverty identification index system of sample villages in the study area. See Table 2.

2.3   Distribution of Key Poverty-Stricken Counties in the Area

As shown in Fig. 1, Research on poverty shows that rural poverty has strong agglomeration in space. According to the spatial distribution of poverty-stricken counties in China, Xiaojin County and Heishui County in Sichuan Province are separated from other poverty-stricken counties. Therefore, the two counties are non-poverty counties, which can't drive their development in a relatively closed environment. According to the national geographic data, the geographical environment of the two counties is similar, with mountainous terrain and high altitude. It can be seen from Table 3 that the disposable income of rural population in the two counties is not much different. Among them, the per capita GDP of Xiaojin County is lower and involves more rural population. To sum up, the research object of this study is Xiaojin County. In this way, it is convenient to guide the mountainous areas with complex terrain and concentrated ethnic minorities and explore the path of self-development in relatively closed areas [8–10].

Table 2 Weight of multi-dimensional poverty indicators

Dimensionality	The dimension weight	Index	Index weight
Financial capital	0.155	Per capita net income (−)	0.248
		Proportion of expenditure on education (−)	0.425
		Share of Medical expenditure (+)	0.327
Natural capital	0.124	Crop area per capita (−)	0.322
		Per capita cash crop area (−)	0.266
		Output of natural resources per capita (−)	0.412
Human capital	0.126	Proportion of population aged 15–19 (+)	0.1
		Proportion of population aged 20–59 (−)	0.136
		Literacy rate (+)	0.163
		Proportion of tertiary education population (−)	0.082
		Proportion of non-agricultural working population (−)	0.1
		Proportion of the population with No ability to work (+)	0.099
		Unable to create economic population ratio (+)	0.155
		Proportion of population with skills training (−)	0.079
		Proportion of party members (−)	0.086
Social capital	0.143	Ratio of per capita net income to minimum wage (−)	0.136
		Type of help received in times of difficulty (−)	0.184
		Level of Trust and communication between Friends and family (−)	0.197
		Number of friends and relatives who keep in touch (−)	0.269

(continued)

Table 2 (continued)

Dimensionality	The dimension weight	Index	Index weight
		Number of friends and relatives who can provide employment information (−)	0.214
Physical capital	0.131	Distance from the county (car) (+)	0.148
		Average housing age (+)	0.208
		Average livestock value (−)	0.184
		Value of household living facilities (−)	0.102
		Per capita electricity consumption (−)	0.13
		Broadband penetration (−)	0.228
Environment/background vulnerability	0.188	Altitude (+)	0.116
		Proportion of slope area above 15° (+)	0.34
		100 + Percentage of long-lived population (−)	0.544

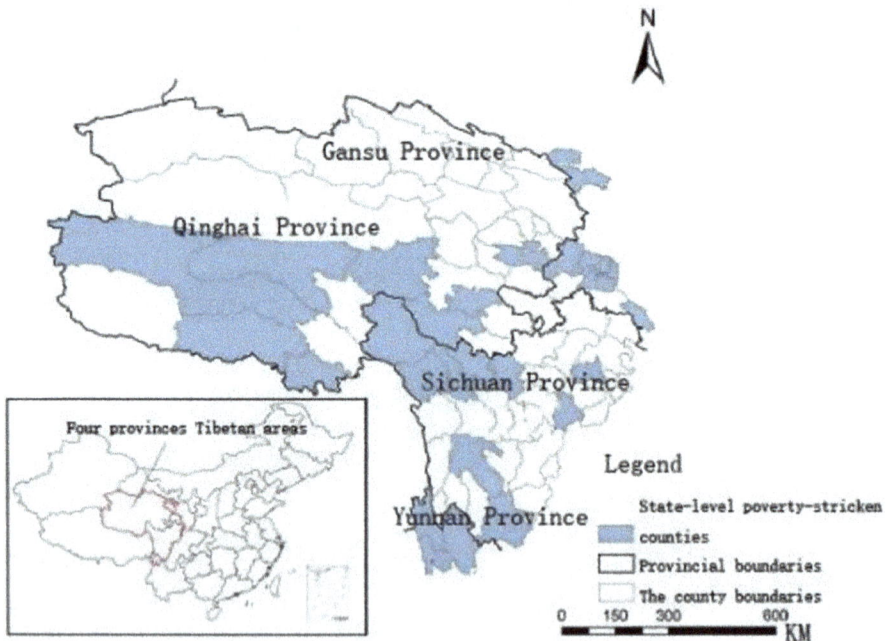

Fig. 1 Distribution of national poverty-stricken counties in Tibetan areas of four provinces

Table 3 Basic situation of key counties with non-centralized poverty alleviation work

	Per capita GRP(YUAN)	Total population (10,000)	Proportion of rural population (%)	Disposable Income of rural residents (YUAN)	Urbanization rate (%)
Xiaojin County	18,112	7.99	64.3	11,596	36.89
Heishui County	34,999	6.05	60	11,515	35.66

2.4 Analysis of Poverty-Stricken Villages in Xiaojin County

The poverty-stricken villages in Xiaojin County are geographically distributed like veins, and most of them are located near the mountains at an altitude of about 3,000 m. Due to the extremely high altitude, there are very few varieties of arable crops. Furthermore, the geological environment of the village was damaged by the "5.12" earthquake, which made the soil loose and the surface rocks piled up, making it extremely difficult to reclaim. It can be seen from Fig. 2 and the nuclear density analysis chart of poverty-stricken villages that the poverty-stricken villages in Xiaojin County are highly concentrated in the southwest with low altitude, which is the same as the altitude when they quit the poverty-stricken administrative villages in 2017. In view of the purpose of exploring regional coordinated development in this study, the selected study area should try its best to include the administrative villages that have

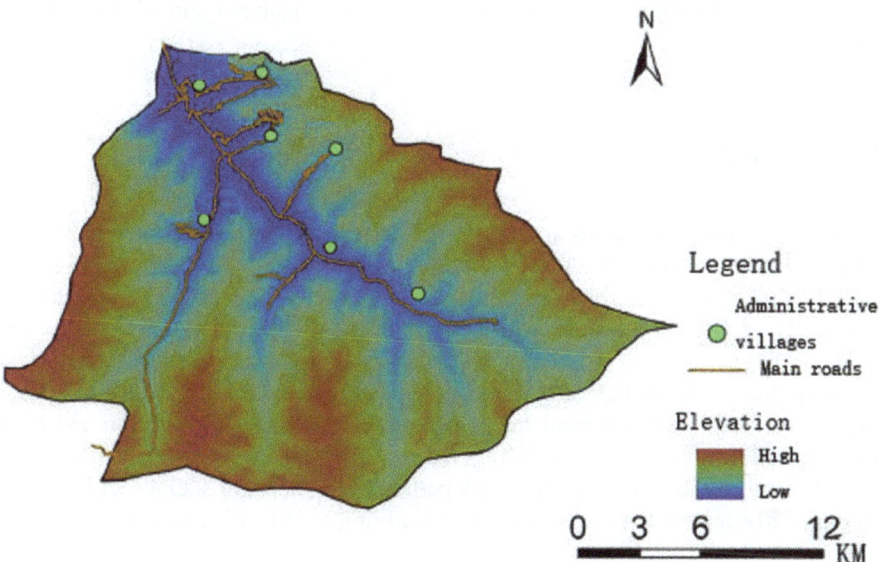

Fig. 2 Spatial distribution map of poor villages in Xiaojin County

been lifted out of poverty, that is, select the study area with moderate concentration of poor villages. The reason is that it plays the role of driving and learning from regional coordinated development. If we explore the coordinated development path of too dense poor villages, it will result in more investment and higher failure rate. By considering the possibility of data acquisition, cost estimation, data analysis, and other factors, this study selected five administrative villages in Meiwo Township. The study area includes three poverty-stricken villages and two poverty-stricken villages, all of which are higher than 3000 m above sea level.

3   Countermeasures

Combined with the multi-dimensional poverty situation and poverty alleviation performance in the study area, the following coordinated development paths are formulated.

Talent revitalization: By protecting local high-quality talents and introducing foreign professionals, the goals of regional coordinated development, characteristic industry construction and cultural heritage description are achieved.

Land consolidation and functional zoning: The main content of land consolidation is land intensification. The land available for planting and development in the study area is limited, and intensive and efficient use of land is imminent. In addition, land intensification across administrative villages not only enhances the comprehensive carrying capacity of the study area, but also enhances its comprehensive competitiveness. Clarifying the functional zoning of land use is conducive to efficient land use, reduces the difficulty of regional planning, and ensures regional integrity and villagers' unity.

4   Conclusion

At present, China's social security system is in urgent need of improvement on the road of building a well-off society in an all-round way, and the rural social security system is the main area for improvement. Therefore, it is necessary to establish a cooperative governance framework of multi-domain subjects against poverty and make a good connection between social security policies and precise poverty alleviation policies. The policy direction should be adjusted to realize multi-pillar rural old-age security, establish rural medical security combining prevention and treatment, introduce active work welfare system, and establish integrated rural social security system. So as to explore a new path, realize the rural social security to get rid of poverty, and realize the grand goal of building a well-off society in an all-round way.

References

1. Pan, D.: The Impact of agricultural extension on farmer nutrient management behavior in Chinese rice production: a household-level analysis. Sustainability **6**(10), 6644–6665 (2014)
2. Tieguhong, J.C., Ndoye, O., Grouwels, S., Mala, W.A., Betti, J.L.: Rural enterprise development for poverty alleviation based on non-wood forest products in Central Africa. Int. For. Rev. **14**(3), 363–379 (2010)
3. Zhang, H., Xu, Z., Wu, K., Zhou, D., Wei, G.: Multi-dimensional poverty measurement for photovoltaic poverty alleviation areas: evidence from pilot counties in China. J. Cleaner Prod. 241:118382 (2019)
4. Tieguhong, J.C., Ndoye, O., Grouwels, S., Mala, W.A., Betti, J.L.: Rural enterprise development for poverty alleviation based on non-wood forest products in Central Africa. Int. For. Revi. **14**(3), 363–379 (2012)
5. Preker, A.S., Carrin, G., Dror, D., Jakab, M., Hsiao, W., Arhintenkorang, D.: Eficacia del financiamiento comunitario de la salud para hacer frente al costo de las enfermedades. Bull. World Health Organ. **2002**(80), 12341–12345 (2002)
6. Liu, N., Li, X., Waddington, S.R.: Soil and fertilizer constraints to wheat and rice production and their alleviation in six intensive cereal-based farming systems of the Indian sub-continent and China. Food Security **6**(5), 629–643 (2014)
7. Quintos, J.R., García, J.A.E., de Pablos Heredero, C., Rivas, J., Perea, J., Angón, E., Martínez, A.G.: Is the increase of scale in the tropics a pathway to smallholders? Dimension and ecological zone effect on the mixed crop-livestock farms. Span. J. Agric. Res. **15**(2), 145–150 (2017)
8. Raab, A.R.T.: Knowledge sharing and distance learning for sustainable agriculture in the Asia-Pacific region: the role of the Internet. Plant Prod. Science **8**(3), 298–307 (2012)
9. Deb, A.K., Haque, C.E.: Sufferings start from the mothers' Womb': vulnerabilities and Livelihood war of the small-scale fishers of Bangladesh. Sustainability **3**(12), 2500–2527 (2011)
10. Valderrama, D., Cai, J.N., Hishamunda, N., Ridler, N., Neish, I.C., Hurtado, A.Q., Msuya, F.E., Krishnan, M., Narayanakumar, R., Kronen, M., Robledo, D., Gasca-Leyva, E., Fraga, J.: The economics of Kappaphycus seaweed cultivation in developing countries: a comparative analysis of farming systems. Aquac. Econ. Manag. **19**(2), 251–277 (2015)

Sensitivity Analysis of NVH Structure of Car Body Based on Modal Analysis

Yongqiang Liu, Xin Liu, Lin Dou, Yu Luan, Lei Shi, and Guolong Zheng

Abstract Taking a certain brand of car as an example, a finite element model of the sound cavity of the body and the passenger compartment was established. By combining the modal analysis with the acoustic structure coupling of the cabin and the passenger compartment, the sound pressure response is used to analyze the coupled vibration mode of the cabin structure and the indoor acoustic cavity, and the indoor noise level is analyzed. The main sources of internal noise are analyzed through structural sensitivity identification. In addition, this article proposes an improved scheme to effectively reduce the noise inside the vehicle. This article is based on a household SUV. Through modal analysis and dynamic frequency response analysis, the internal noise caused by structural vibration is studied, and suggestions for improving the body parts are put forward to reduce the internal noise level.

1 Modal Analysis

1.1 Finite Element Modal Analysis of Car Body Structure

The flowchart of sensitivity analysis of car body structure is as follows:

The finite element equation of car body structure can be written as:

$$M_8\ddot{u} + C_8 u + K_8 u = F_8 \tag{1}$$

where M is the structural mass matrix; $\ddot{u}$ is the acceleration of the structural nodes; C_8 is the (structural) damping matrix; iu is the velocity of the structural nodes; K_8

Y. Liu (✉)
State Key Laboratory of Traction Power, Southwest Jiaotong University, Chengdu, China
e-mail: yongqiangliu94@163.com

X. Liu · L. Dou · Y. Luan · L. Shi · G. Zheng
Technology Development Department of FAW-Volkswagen, Changchun 13000, China

© The Author(s), under exclusive license to Springer Nature Singapore Pte Ltd. 2021
L. C. Jain et al. (eds.), *3D Imaging Technologies—Multidimensional Signal Processing and Deep Learning*, Smart Innovation, Systems and Technologies 236,
https://doi.org/10.1007/978-981-16-3180-1_11

Table 1 Stiffness calculation results and experimental results

Performance indicators	The calculation results	The experimental results	Error (%)
The bending stiffness	7852.84 N/mm	7492.62 N/mm	4.81
Torsional rigidity	7367.45 N°m/(°)	7086.33 N°m/(°)	3.96

is the structural stiffness matrix; u is the displacement of the structural nodes; F_8. is the External forces applied to the structure.

The body model is established in Altair. Hypermesh. The body $M_8\ddot{u} + C_8 u + K_8 u = F_8$ adopts triangle and quadrilateral grid elements (the finite element model of the body is composed of triangle and quadrilateral grid elements), the total number of elements is 347,105, and the percentage of triangle cells (the ratio of triangle grid elements to the total number of elements) is 6%. A cWELD unit was used for body welds, totaling 5851, and a CBAR unit was used for bolts. The stiffness of the body model is compared with the experimental results, and the error is small, which meets the requirements of analysis. The calculated and experimental results of stiffness are given in Table 1 [1–3].

Modal parameters of automobile structure reflect the inherent vibration characteristics of automobile structure, and automobile structure has an important influence on the internal noise of automobile and also affects the stability of automobile running. For example, torsion (one of vibration forms, which is reflected in structural mode) is one of the most influential vibration forms on the structure. The first-order torsional mode of vehicle body is an important parameter in automobile design, which represents the torsional resistance of structure at low frequency resonance. In the design, the vibration mode frequency (natural mode frequency of structure) with great harm should be staggered with the excitation frequency of road surface and engine as much as possible [4–6].

1.2 Finite Element Modal Analysis of Acoustic Cavity

In the calculation of acoustic fluid, it is assumed that the fluid can be properly compressed, without flow and viscosity, so that the cabin sound pressure satisfies Helmholts wave equation:

$$\frac{1}{c^2}\frac{\partial^2 p}{\partial t^2} - \nabla^2 p = 0 \tag{2}$$

In the formula, c is the speed of sound in the air; p is the sound pressure; ∇ it is the Laplace operator. Discretize the cabin cavity into finite element, without considering the damping, the sound wave square, the cabin cavity is discretized into finite element, and the acoustic wave equation can be written as:

$$M_f \ddot{p} + K_f p = F_g \tag{3}$$

In the formula, M_f is the cavity mass matrix; $\ddot{p}$ is the second-order reciprocal of the cavity sound pressure versus time; K_f is the cavity stiffness matrix; p is the cavity sound pressure; F_g is the generalized force transmitted to the fluid by the structural unit.

The modal frequencies and modes of acoustic cavity are shown in Fig. 2. For convenience of explanation, the terms "longitudinal," "transverse," and "vertical" are introduced, which correspond to the X, Y, Z, Y and Z axes of the vehicle coordinate system, respectively. Acoustic resonance frequency is very important in car design. We can see that the black pitch line position (zero sound pressure position) in Fig. 2 is located in the middle and front of the car room, which enables people to be in the acoustic environment with the least noise (the minimum sound pressure at this time is 41.84 mPa).

1.3 Modal Analysis of Acoustic-Solid Coupling

The indoor sound pressure distribution of the passenger compartment is calculated in a closed car body structure, and the front and rear windshields, window glass and car body (metal car body composed of welded steel plates) are connected, and 938,970 units of the acoustic-solid coupling model are stitched by hand, as shown by the frequency and mode shape of the first-order coupling mode sound cavity. Coupling modes include not only the distribution of sound pressure in the cabin cavity, but also the change modes of the structure, among which some are mainly the change of fluid sound pressure and the other are mainly the deformation of the structure. The mode shape and frequency of acoustic cavity mode have changed after coupling. In the first mode shape, the pitch line position (zero sound pressure position) is located at the driver, which can effectively ensure the minimum noise at the driver (the minimum sound pressure value is 4.58 Pa at this time) [7–9].

2 Sound Pressure Response Analysis

The response of the system to broadband excitation can be obtained by exciting the system with white noise excitation force within a certain frequency range, that is, the noise level in the cabin under different working frequencies can be obtained. Given the white noise unit excitation of 10–200 Hz, it is loaded at four suspension points, respectively. According to the relevant standards, the node at the driver's right ear is selected as the reference point, and the frequency noise curve here is obtained. There are resonance peaks 1 and 2 at 65 Hz and 90 Hz, in which the sound pressure value at 65 Hz is 4.7 Pa and the sound pressure value at 90 Hz is 1.37 Pa, which shows that when the car body structure encounters the excitation of these two frequencies,

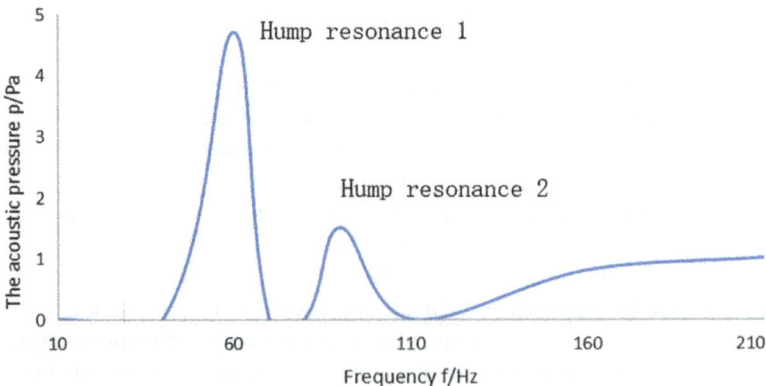

Fig. 1 Sound pressure–frequency curve of driver's reference point

Fig. 2 Sound pressure–frequency curve of driver reference point after improvement

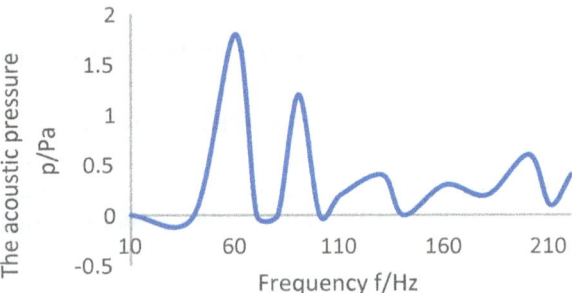

it is easy to produce obvious noise in the cabin, which is obviously not conducive to improving the ride comfort. It is helpful for noise control to determine the influence degree of the components of the car body structure on these two frequencies. In this example, the influence of seat space and interior sound absorption is not considered, so the internal sound pressure value is higher than the real value. See Fig. 1.

3   Structural Sensitivity Analysis

The medium and low frequency noise in the car is mainly related to the vibration of the car body structure, and the sensitivity of different panels of the car body to the sound pressure at a certain point in the interior space of the passenger compartment is also different. Therefore, accurately determining the relationship between structural vibration and vehicle interior noise can effectively suppress the structural vibration of the vehicle body and achieve the purpose of controlling the vehicle interior noise.

The cabin panel surrounding the acoustic cavity is divided into eight parts: front firewall, front windshield, front door, front floor, rear door, rear cover door, rear floor, and ceiling. The node sets are filled in the solver card according to the format, and the sensitivity values of each panel to the sound pressure points in the car can be obtained through the sensitivity analysis of the car body. In order to verify the proposed solution, through the analysis results of panel sensitivity, the vehicle interior noise sensitivity is modified to obtain the thickness of the vehicle body parts. Considering the current steel plate specification enterprises, the thickness of the inner and outer plates of the rear door, the inner and outer plates of the front door and the front floor is reduced by 0.2 mm. Increase the thickness of the windshield by 0.6 mm and the thickness of the ceiling to 0.3 mm, and analyze the sound pressure again accordingly. The results are shown in Fig. 2. At 65 and 90 Hz, the sound pressure decreased significantly, among which the sound pressure at 65 Hz decreased to 1.74 Pa, which was 8.6 db lower in decibel, and the sound pressure at 90 Hz decreased to 0.64 Pa, which was 6.6 dB lower in decibel. At the same time, the sound pressure at other frequencies increased slightly, among which the sound pressure at 125 Hz increased from 0.13 to 0.32 Pa, and the sound pressure at 175 Hz increased from 0.23 to 0.55 Pa, but it was still significantly smaller than the sound pressure at the first two resonance peaks. This shows that the acoustic pressure values of other frequencies will be affected by the thickness of several modified parts, but the sensitivity of these parts at frequencies is not high, which leads to a slight increase in the acoustic pressure values at these modified frequencies, but it is still significantly smaller than the resonance peak before improvement. Through the improvement, the sound pressure at the two resonance peaks in the car is reduced, and the purpose of reducing the noise in the car is achieved, which effectively verifies the implementation of this scheme (Fig. 2).

4 Conclusion

In this paper, taking a domestic SUV as an example, the finite element models of body structure and sound cavity of passenger compartment are established respectively, and the finite element model of sound-structure coupling system is established on this basis. Based on the modal analysis method, the modal characteristics of the car body structure and the sound cavity of the passenger compartment are analyzed, respectively, and the modal and vibration characteristics of the sound-structure coupling system are analyzed. Through the analysis of sound pressure response, the noise inside the car and its frequency distribution characteristics are given, and two noise peaks are obtained. Through the structural sensitivity analysis of peak frequency, the sensitivity coefficient of car body parts to noise peak is determined, the parts with higher sensitivity are found out, and the improvement measures are put forward. The calculation results show that the two noise peaks have dropped obviously, which verifies the correctness of the analysis method.

Acknowledgements This project is supported by National Natural Science Foundation of China (Grant No. U1934202).

This project is supported by National Natural Science Foundation of China (Grant No. 51975485).

References

1. Kropp, A., Heiserer, D.: Efficient Broadband vibro-acoustic analysis of passenger car bodies using an FE-based component mode synthesis approach. J. Comput. Acoust. **11**(02), 139–157 (2011)
2. Gmachowski, L., Iwata, M., Adachi, T., Murase, T.: Analysis of floc structure based on the extended Brinkman model. J. Chem. Eng. Jpn **23**(5), 543–549 (1990)
3. Zemlyanaya, E.V., Kiselev, M.A., Zbytovska, J., Almasy, L., Aswal, V.K., Strunz, P., Wartewig, S., Neubert, R.: Structure of unilamellar vesicles: numerical analysis based on small-angle neutron scattering data. Crystallogr. Rep. **51**(S1), S22–S26 (2006)
4. Jia, D., Li, F., Zhang, C., Li, L.: Design and simulation analysis of trimaran bulkhead based on topological optimization. Ocean Eng. **191**(1), 106304.1–106304.39 (2019)
5. Edwards, M.J., Fredrickson, J.K., Zachara, J.M., Richardson, D.J., Clarke, T.A.: Analysis of structural MtrC models based on homology with the crystal structure of MtrF. Biochem. Soc. Trans. **40**(6), 1181–1185 (2012)
6. Nekrasov, A.A., Ivanov, V.F., Vannikov, A.V.: Analysis of the structure of polyaniline absorption spectra based on spectroelectrochemical data. J. Electroanal. Chem. **482**(1), 11–17 (2000)
7. Kobryansky, V.M., Tereshko, E.A.: Analysis of the electronic structure of polyacetylene based on its optical absorption spectra. Synth. Metals **39**(3), 367–378 (1991)
8. Marcinkowski, M.J., Fisher, R.M.: Theoretical analysis of plastic deformation in superlattices based on the body-centered cubic structure. J. Appl. Phys. **34**(8), 2135–2145 (1963)
9. Zeng, H.C., Peltola, H., Talkkari, A., Strandman, H., Venäläinen, A., Wang, K., Kellomäki, S.: Simulations of the influence of clear-cutting on the risk of wind damage on a regional scale over a 20-year period. Can. J. For. Res. **36**(9), 2247–2258 (2006)

Research on the Design of Nutrition and Sports Health Management System Based on Artificial Intelligence

Bolun Wang

Abstract In recent years, in the wave of big data on the Internet, China's artificial intelligence development has always been among the top in the world. With the popularization of big data, the Internet and artificial intelligence have been involved in various fields, and the combination of artificial intelligence and current health management has achieved remarkable results. In this case, this paper advocates to provide differentiated health management services with the support of relevant literature and discusses the necessity of developing artificial intelligence in professional health management. Explore the development and design concept of personalized sports management service based on the concept of artificial intelligence, system operation process, management stage development, etc. Researched the practical ways of national health management. Artificial intelligence has a profound impact on promoting the expansion and renewal of sports and health management services, improving the level of sports and health management and opening up national fitness.

1 Questions

In recent years, the aging of the population trend, young people with cancer, sub-health, these prominent health problems have become a new concern of people and society. Although the enlightenment of artificial intelligence in our country is late, with the leading e-commerce companies and related electronic industries conducting massive layout and research and development, the relative investment and entrepreneurial passion in the industry are stimulated. Now, artificial intelligence lies in the sports health management service theory applied to explore the still weak, but now the development of the society and people's needs are urgent need sports health management service, so now for the study of personalized management system under the artificial intelligence development update, help improve physical health, open the health, it makes the development of health of China constantly [1–4].

B. Wang (✉)
School of Communication, Hong Kong Baptist University, Hong Kong, China

© The Author(s), under exclusive license to Springer Nature Singapore Pte Ltd. 2021   91
L. C. Jain et al. (eds.), *3D Imaging Technologies—Multidimensional Signal Processing and Deep Learning*, Smart Innovation, Systems and Technologies 236,
https://doi.org/10.1007/978-981-16-3180-1_12

2 Personalized Sports Management

From the beginning of foreign scholars' exploration of personalized health management, to the continuous renewal and steady development of the present, we have been constantly exploring and innovating. Under the application of Internet and new media technology, the future direction of exploration is more inclined to the practice of personalized health services and online electronic records of health records. China is still in the enlightenment stage in this field, and the research subject cases and appeal are not as good as those of foreign countries. In the medical field, radiotherapy and treatment are the main approaches, and the relevant prevention research practices are insufficient. At the 2010 Summit forum, the vice President of the Ministry of Health proposed the health project of "preventing diseases," established the preventive service system of health care, and improved the corresponding norms and models. Therefore, there are still great opportunities and challenges for the independent development of personalized health management in China, which requires us to increase the support of science and technology, human resources, and material resources in this service field [5–7].

3 The Necessity of Personalized Sports Health Management Service in the Era of Artificial Intelligence

3.1 *Intelligent Products Provide Scientific Data for Personalized Sports and Health Management Services*

In recent years, our country has developed and expanded rapidly in the field of artificial intelligence, and the artificial intelligence market and related industries have undergone "fission" growth. According to the development trend, 33.56 billion yuan will be created by artificial intelligence in 2022. Artificial intelligence is involved in nearly 20 industries, such as schools, hotels, banks, malls, museums, and so on. In the future, it will expand its territory and apply in other fields. At the present stage, smart bracelets, shoes, shirts, and glasses are all put into practice with the support of smart technology, and the above products will become indispensable products for residents' life. It can monitor the health data being maintained at any time and reflect the effects and problems, helping residents to understand how to improve their health [8–11].

3.2 Intelligent Stadiums Optimize the Sports Environment for Personalized Sports and Health Management

With the continuous development of today's society, intelligent stadium is born with the application, which provides a characteristic sports health management and service environment for sports. People in the scene of intelligent movement, the corresponding services, are also constantly improving. It has created an efficient sports environment for the people and formed a benign business model. To provide the public with detailed and characteristic fitness guidance, reflects the organic combination of scientific sports and public health [12].

4 Design of Personalized Sports and Health Management Service System in the Era of Artificial Intelligence

4.1 Design Ideas of Artificial Intelligence + Personalized Sports Management Service System

At present, the intelligent sports management mode is based on the "monitoring, evaluation, intervention, and promotion" of sports and health management service to construct the human intelligent personalized sports and health management service system.

4.2 Operation Process of Artificial Intelligence + Personalized Sports Management Service System

As can be seen from Fig. 1, the operation process of the "artificial intelligence + personalized sports management service" system is relatively complex and tedious. The system overturns the traditional sports health management mode and builds a new sports health management system based on the Internet of Things and cloud computing.

4.3 Design of Artificial Intelligence + Personalized Sports Management Service Management Stage

Sports health management service links are divided into six links: sports health information collection system, sports health status assessment and prediction, sports health

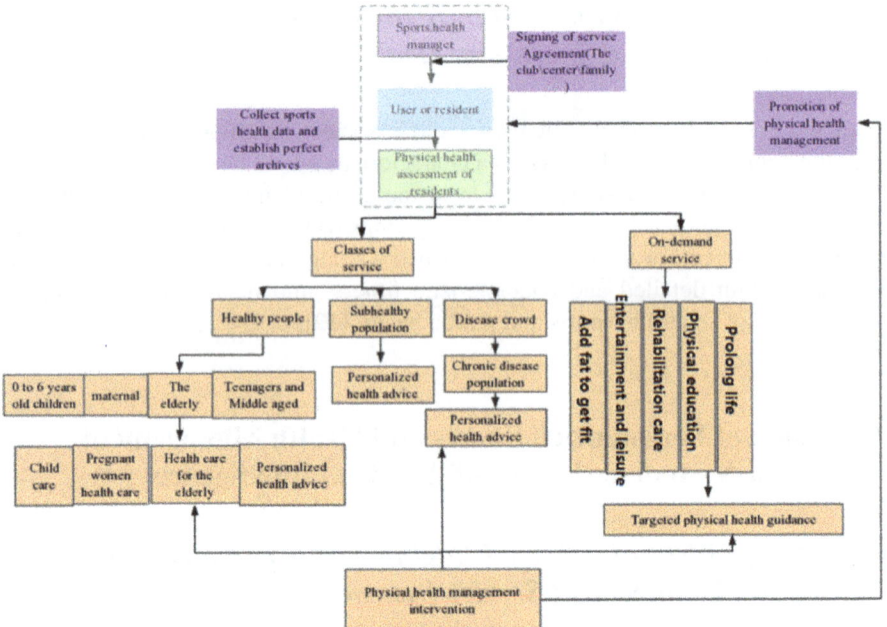

Fig. 1 Operation flow chart of "Artificial Intelligence + Personalized Sports Management Service" system

management plan formulation, sports health intervention, sports health promotion and sports health assessment.

5 Implementation Path of Personalized Sports and Health Management Services in the Era of Artificial Intelligence

5.1 Implementation Path of Self-Help Sports and Health Service Based on Family or Hospital

At present, the resident health management information platform built by the third-party medical operation institutions provide remote sports and health management information operation services to family members as a unit, forming the application system of self-service sports and health management services in the family physical therapy in the hospital.

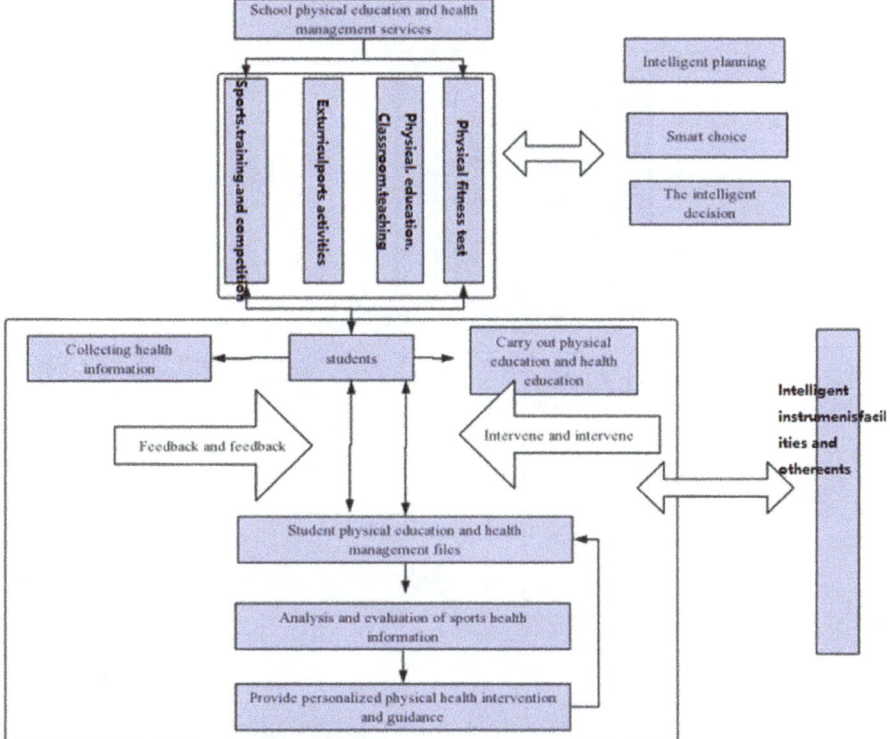

Fig. 2 Implementation path diagram of school physical education and health service management

5.2 Implementation Path of Physical Education and Health Management Service Based on Schools

It can be seen from Fig. 2 that, under the unified leadership of the school, the physical education and research office of the school USES intelligent equipment or instruments to carry out physical health tests, physical education classroom teaching, extracurricular physical activities, sports training, and competitions as the main contents of the health management system.

5.3 Implementation Path of Individualized Sports and Health Management Service Based on Community

From Fig. 3 shows, community sports health management service in the city, and under the leadership of the municipal district people's government, dominated by

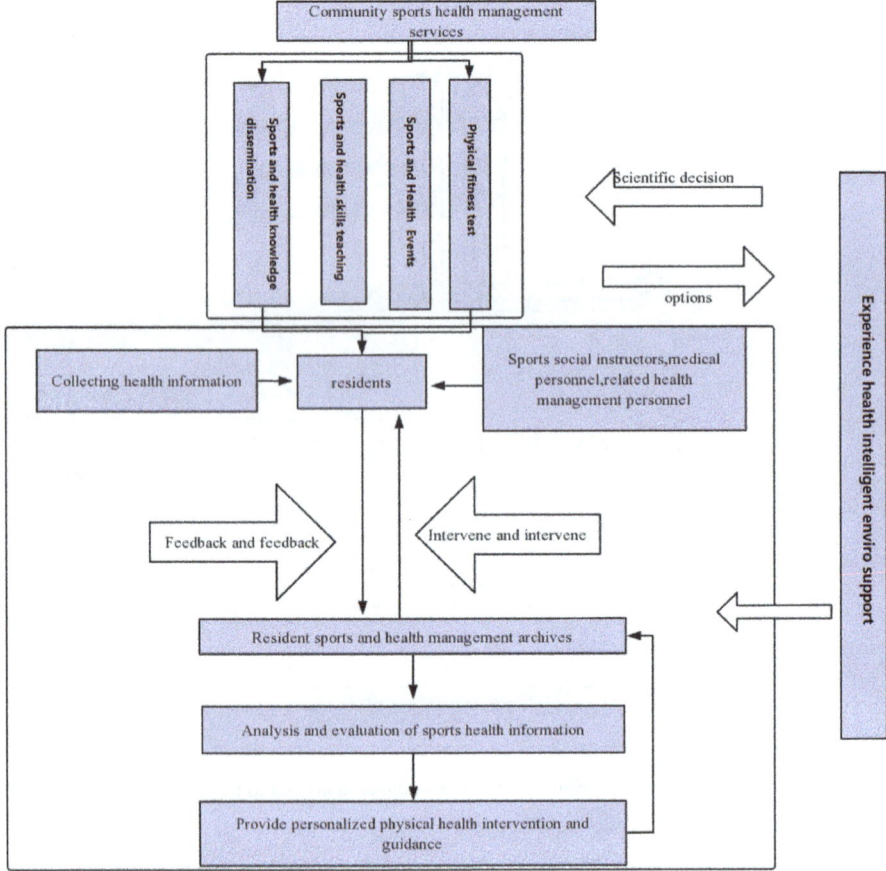

Fig. 3 Implementation path diagram of community sports health service management

street, neighborhood committees for coordination, establish a sports health management team as a guide, geared to the needs of community residents to carry out physical health test, sports health events, sports skill teaching, sports health knowledge dissemination as the main content of the management system.

6 Conclusion

China's sports health management services have been artificial intelligence products and technology penetration, but not deep enough. Artificial intelligence applied to sports health management service theory research is very weak, social development, and the urgent need of sports health management services, so the artificial intelligence under the age of personalized sports health management service system design, for power transformation and upgrading of sports health management services, and boost the sports health management service supply of innovation and development, promote the national fitness and health and provide reference basis for the healthy development of our country.

References

1. Li, C., Wang, Z.: Research on the Applications of Information Technology in Sport Management. In International Conference on Information and Business Intelligence, pp. 247–252. Springer, Berlin, Heidelberg (2011)
2. Dittenbach, M., Rauber, A., Merkl, D.: Business, culture, politics, and sports how to find your way through a bulk of news? On content-based hierarchical structuring and organization of large document archives. In: International Conference on Database and Expert Systems Applications, pp. 200–210. Springer, Berlin, Heidelberg (2001)
3. Xu, X., Yang, J., Xu, H., Tan, Y.: Towards the representation of virtual gymnasia based on multi-dimensional information integration. In: Chinese Conference on Image and Graphics Technologies, pp. 249–256, Springer, Berlin, Heidelberg (2014)
4. Wei, W.: The Research on the Psychological Quality for University Sports Development Training. In Emerging Computation and Information Technologies for Education, pp. 321–327, Springer, Berlin, Heidelberg. (2012)
5. Incel, O.D., Kose, M., Ersoy, C.: A review and taxonomy of activity recognition on mobile phones. Bionanoence **3**(2), 145–171 (2013)
6. Kittsteiner, T., Ockenfels, A.: On the design of simple multi-unit online auctions. In: Negotiation. Auctions, and Market Engineering, pp. 68–71. Springer, Berlin, Heidelberg (2008)
7. Bai, H., Hu, W., Wang, T., Tong, X., Zhang, Y.: A novel sports video logo detector based on motion analysis. In: International Conference on Neural Information Processing, pp. 448–457, Springer, Berlin, Heidelberg (2006)
8. Lang, C., Xu, D., Cheng, W., Jiang, Y.: Shot Type Classification in Sports Video Using Fuzzy Information Granular. In International Conference on Knowledge-Based and Intelligent Information and Engineering Systems, pp. 1217–1223, Springer, Berlin, Heidelberg (2005)
9. Pan, Z., He, G., Su, S., Li, X., Pan, J.: Virtual network marathon: Fitness-oriented E-Sports in distributed virtual environment. In International Conference on Virtual Systems and Multimedia , pp. 520–529, Springer, Berlin, Heidelberg (2006)
10. Macedo, P., Madeira, R. N., Correia, A., Jardim, M.: A Web System Based on a Sports Injuries Model towards Global Athletes Monitoring. In: New Perspectives in Information Systems and Technologies, vol. 2, pp. 377–383, Springer, Cham (2014)
11. Wang, D., Han, Y., Jiao, A.: The research on supplement of glutamine in sports. Adv. Intell. Soft Comput. **4**(24), 241–243 (2012)
12. Brzostowski, K., Drapała, J., Dziedzic, G., Świątek, J.: Algorithm to plan athlete's prolonged training based on model of physiological response. In: Asian Conference on Intelligent Information and Database Systems, pp. 273–283, Springer, Cham (2015)

Research on English Word Intelligent Learning System Based on Mobile Education Concept

Ting Deng

Abstract Because traditional education is easily restricted by learning place and time, it is difficult to meet people's learning needs. With the rapid development of information technology, information technology has been widely applied to the education industry, resulting in a new mobile education model. Mobile education is an educational way in which users use mobile devices such as mobile phones to educate and learn. At present, the government vigorously implements education reform, and mobile education has become the main method to improve the level of education for all, and it is also the focus of future education development. Mobile education is a brand-new form of distance learning. Users can get rid of the constraints of time and space and receive education anytime and anywhere. Based on the advantages of the development mode system in the field of English instructional design, this paper puts forward the development mode of intelligent ubiquitous learning system for English words, so that English word learners are not limited by time and space, and provide reference for the design and development of mobile learning software system.

1 Introduction

Distance education in China has experienced correspondence education, TV broadcasting education, and network education with the help of modern information technology, which has trained a large number of outstanding talents for our country [1]. The evolution of distance education is closely related to the development of media. With the development of media technology, educators can jump out of the traditional education mode of simultaneous, local, and synchronous education, which is not limited by time and place. Under the digital environment, the change of learning style not only improves the efficiency of education, but also sets off an upsurge of educational research. With the popularization of mobile communication terminals and the innovation of wireless communication technology, mobile education came into being [2]. Mobile education has become the focus of people's attention because

T. Deng (✉)
JiangXi Teachers College, Yingtan 335000, Jiangxi, China

it makes up for the deficiency of traditional education in teaching mode. Mobile education belongs to the new development field of educational technology and has broad prospects. By studying the related theories and application methods in the field of mobile education, the development of mobile education can be effectively promoted. The mobile learning system can not only exert the processing ability of the client, but also satisfy the real-time network service. When students' English level reaches a certain level, they usually stagnate and cannot break through innovation. Vocabulary is usually regarded as an important factor. Students' limited vocabulary can directly affect their reading ability, listening comprehension ability, oral expression ability, and written expression ability, and even improve their overall English level. Therefore, it is an urgent problem to help students break through the "bottleneck" and effectively expand their vocabulary. English words intelligent ubiquitous learning system based on mobile education support platform can provide learners with English words ubiquitous for learning. Learners can use the platform to study anytime and anywhere, so as to help students use their spare time, break through the "bottleneck" and effectively expand their vocabulary [3, 4].

2 System Analysis and Design

The development mode of intelligent ubiquitous learning system for English words is as follows:

A key problem of this system is to select words intelligently from thesaurus. Intelligent word selection is to search for words that meet the requirements of teaching and readers in thesaurus, which is an important subject in modern intelligent teaching system. In this subject, genetic algorithm, as a global optimization search algorithm, has been widely used. It can select more adaptable individuals from the group according to the laws of nature, so as to get the optimal solution [5, 6]. It has intelligent characteristics such as self-organization, self-adaptation, and self-learning. Example: let the value range of a certain parameter be $[U_1, U_2]$ (Fig. 1).

We use the binary code symbol with length k to represent the parameter, so it produces a total of 2^k different codes, which can make the corresponding relationship when the parameter is coded:

$$
\begin{aligned}
000000 \cdots 0000 &= 0 \rightarrow U_1 \\
000000 \cdots 0001 &= 1 \rightarrow U_1 + \delta \\
000000 \cdots 0010 &= 2 \rightarrow U_1 + 2\delta
\end{aligned}
\tag{1}
$$

In which:

$$
\delta = \frac{U_2 - U_1}{2^k - 1}
\tag{2}
$$

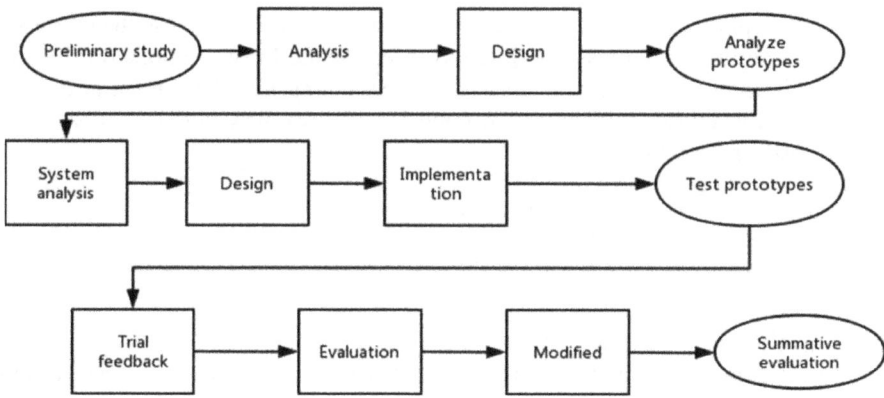

Fig. 1 Development mode of learning system

Assuming the coding of a certain volume, the corresponding decoding formula is:

$$X = U_1 + \left(\sum_{i=1}^{k} b_i \cdot 2^{i-1} \right) \cdot \frac{U_2 - U_1}{2^k - 1} \tag{3}$$

X is encoded with 5-bit binary code, and the following binary string is obtained:

00000,00001,00010,00011,00100,00101,00110,00111
01000,01001,01010,01011,01100,01101,01110,01111
10000,10001,10010,10011,10100,10101,10110,10111

For any binary system, as long as the above formula is brought in, the corresponding decoding can be obtained. According to the relative fitness selection probability, the standard turntable selection strategy in genetic algorithm is used to select the new group of individuals. The fitness of relative fitness is defined as follows, popsize is the population size, and fi is the set function.

$$\text{fitness} = \frac{f_i}{\sum_{i=0}^{popsize} f_i} \tag{4}$$

In the running process of genetic algorithm, new individuals are constantly produced by genetic operations such as crossover and mutation. Although more and more excellent individuals will be produced with the evolution of the population, they may also destroy the best fitness in the current population because of the randomness of genetic operations such as selection, crossover, and mutation.

The individual. Therefore, it is hoped that the individuals with the best fitness will be kept in the next generation as much as possible. To achieve this goal, we can use the optimal preservation strategy evolution model, that is, use the crossover method to survive the fittest [7–9]. Cross refers to the linear combination of two individuals to produce two new individuals. Assuming that there is an arithmetic crossover between two individuals, *Xat* and *Xbt*, the two new individuals produced after crossover operation are:

$$\begin{cases} X_A^{t+1} = \alpha X_B^t + (1 - \alpha) X_A^t \\ X_B^{t+1} = \alpha X_A^t + (1 - \alpha) X_B^t \end{cases} \tag{5}$$

α is the only parameter, and its value range is 0–1.

The screening is repeated until the new group is satisfied, so the intelligent learning system of English words can be configured according to this algorithm.

3 System Implementation

There are two basic use cases in the intelligent ubiquitous learning system of English words: a word learning use case and a word practice use case. Two use cases, "word learning" and "word preview," are used in the process of learners' learning, and these two use cases have the action of searching word bank. Therefore, "searching thesaurus" is regarded as an independent use case. There is also an executor administrator in the system. Administrators use the Update Thesaurus use case. The hardware environment of English word intelligent ubiquitous learning system based on mobile education support platform is mainly composed of four parts as shown in Fig. 2.

Considering the feasibility and overall architecture of the system from the perspective of software engineering. The result of this stage is to obtain the initial prototype of the intelligent ubiquitous learning system of English words and realize the intelligent word selection algorithm (Fig. 3).

Based on the prototype analysis, an intelligent ubiquitous learning system for English words is designed and implemented, and the software code is written. Use Visual C++ for development and debugging. As a result, a relatively complete prototype for trial use is obtained [10].

English word entries are displayed in the text box, and the entries contain English words and word annotations. In the process of learning, it can provide phonetic reading of words and enhance learners' learning efficiency [11].

4 Conclusion

Mobile education is the product of the combination of mobile communication technology, network technology, multimedia technology, and contemporary education,

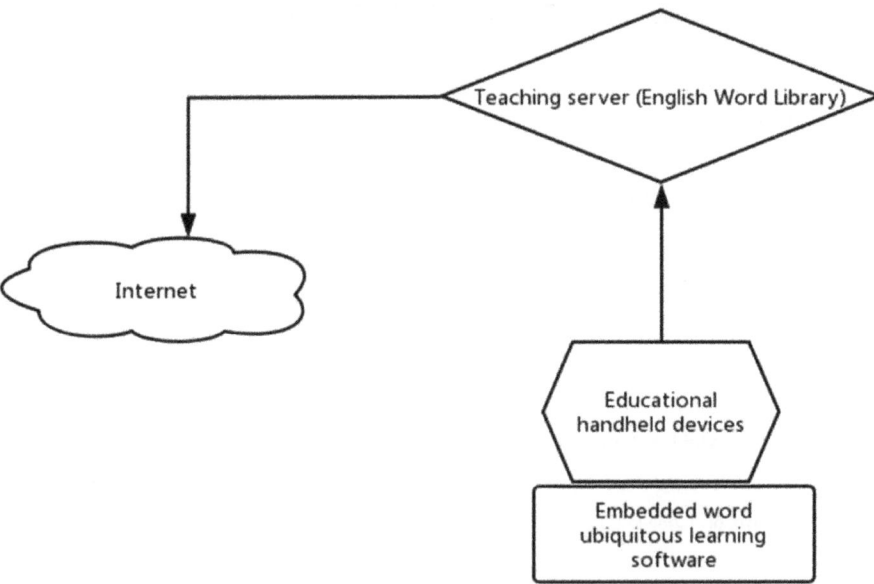

Fig. 2 Hardware environment of intelligent ubiquitous learning system for English words

and it is also an important feature and advantage of modern educational technology. By popularizing and developing mobile education, educational technology and means will be changed. Through the intelligent ubiquitous learning system of English words based on mobile education support platform, this paper explores the development mode of educational software by using handheld mobile terminals such as mobile education learning system and computer, so as to provide a ubiquitous learning platform for English words learning, which can be used by learners to learn English words anytime and anywhere.

```
while(gen<MaxGen&&The target function of the best individual does not meet the require-
ment)

{

for(i=O;i<POPSIZE;i++)

{

Evaluate fitness of P(gen);

}

for(i=0;i<POPSIZE;i++)

{

Select operation to P(gen);

}

for(i=O;i<POPSIZE/2;i++)

{

Crossover operation to P(gen);

}

for(i=O;i<POPSIZE;i++)

{

Mutaton operation to P(gen);

}

for(i=O;i<POPSIZE;i++)

for(i=O;i<POPSIZE;i++)

{

P(gen +1)=P(gen);

}

gen++;

}
```

Fig. 3 Code example

References

1. Zhou, Q.P.: Research of high-voltage switchgear intelligent based on PLC centralized control system. Electr. Eng. **019**(009), 73–76, 81 (2018)
2. Elsobky, M., Mahsereci, Y., Yu, Z., Richter, H., Burghartz, J.N., Keck, J., Klauk, H., Zschieschang, U.: Ultra-thin smart electronic skin based on hybrid system-in-foil concept combining three flexible electronics technologies. Electronics Lett. 54(6):338–340(2018).
3. He, Y., Yang, S., Chan, C.Y., Chen, L., Wu, C.: Visualization analysis of intelligent vehicles research field based on mapping knowledge domain. IEEE Trans. Intell. Transp. Syst. **2**(99), 1–16 (2020)

4. El Hajji, R., Boussetta, A., Aznag, B., Agdouz, F., Ihssane, B., Hamedane, O.A., Saffaj, T.: Comparison of fit for a future purpose concept based on tolerance interval and ISO 11352 approaches to predict the quantitative performances and routine uncertainty of an ICP-MS method for the determination of 21 elements in drinking water. Microchem. J. **138**, 255–265 (2018)
5. Peleg, R., Yayon, M., Katchevich, D., Moria-Shipony, M., Blonder, R.: A lab-based chemical escape room: educational, mobile, and fun! J. Chem. Educ. **96**(5), 955–960 (2019)
6. Peng, D., Xu, Y., Zhao, H.: Research on intelligent predictive AGC of a thermal power unit based on control performance standards. Energies **12**(21), 4073 (2019)
7. Jiang, M., Xu, L., Tao, C., Lu, X.: Research on interaction and evaluation method of learning intelligent system based on knowledge memory. Lect. Notes Electr. Eng. **271**(4), 173–181 (2014)
8. Cui, M.Y., Yang, L., Li, L.J., Zhao, L.B., Yang, F.H.: Research on the development system of intelligent materials based on the concept and technology of large data. Electron. World **9** (2018)
9. Guo, X., Liu, A., Li, X., Xiao, Y.: research on the intelligent fault diagnosis of medical devices based on a DEMATEL-fuzzy concept Lattice. Int. J. Fuzzy Syst. **22**(7), 2369–2384 (2020)
10. Zhang, J.S., Wang, Q.F., Liu, Q.H., Wan, L., Zhong, Y.F.: Research on model-based intelligent product configuration. Chin. J. Mech. Eng. **39**(6), 128–134 (2003)
11. Li, W., Huang, Q.: Research on intelligent avoidance method of shipwreck based on bigdata analysis. Po. Marit. Res. **24**(s3), 116–118 (2017)

Research on the Application of Big Data Analysis in English Language Education

Ting Deng

Abstract With the development of big data technology, it has been widely concerned in the field of English language education and has become the main factor affecting English language education. The reason is that using big data can break the inherent equilibrium of traditional English language education ecosystem. Building an ecological environment for English language education with the help of big data technology can improve the quality of English teaching. This paper studies the relationship among various factors involved in English language education through big data mining technology, so as to build an educational ecosystem oriented to self-adaptive balance adjustment. According to the characteristics of English autonomous learning system in the era of big data, this paper analyzes the problems existing in the current English learning system (such as low level of online and offline collaboration, unsatisfactory combination of instrumental and humanistic effects, and low degree of integration of information technology and curriculum design), constructs English learning environment and language learning situation according to information technology, and puts forward a new English teaching mode combining online and offline, so as to enhance the strategy of guiding students' English autonomous learning and construct a multi-evaluation system of English autonomous learning, so as to build an ecological English language education environment, realize the ecological balance of English language education, effectively improve the quality of English teaching, and provide new ideas for students' English language education in the era of big data.

1   Introduction

With the rapid development of mobile Internet, cloud computing, big data, social network and other networks, and technologies, the integration of information technology and education has become an important force to promote the reform of the education industry. New English teaching models (like flip class, micro-class,

T. Deng (✉)
Jiangxi Teachers College, Yingtan 335000, Jiangxi, China

MOOC, etc.) put forward higher requirements for students' autonomous learning ability. Big data analysis technology provides a new direction for English autonomous learning [1]. Through in-depth study of the characteristics of autonomous learning in the era of big data, students' autonomous learning ability in English can be improved in the new period, and the reform of English teaching mode in education departments can be promoted. Starting with the data mining of students' English learning needs, grades, habits, and preferences, with the help of big data analysis technology, it can effectively combine the massive rich media English teaching content with students' fragmentation time, knowledge and interest of English teaching content, instrumental and humanistic language learning, knowledge memory, and innovative thinking in the learning process, so as to provide information technology support for students' autonomous English learning [2]. The architecture of intelligent English language education in colleges and universities created in this paper belongs to multi-level architecture, which is divided into perception layer, communication layer, cloud computing layer, business layer, data layer, and intelligent service layer. The specific architecture is shown in Fig. 1 [3].

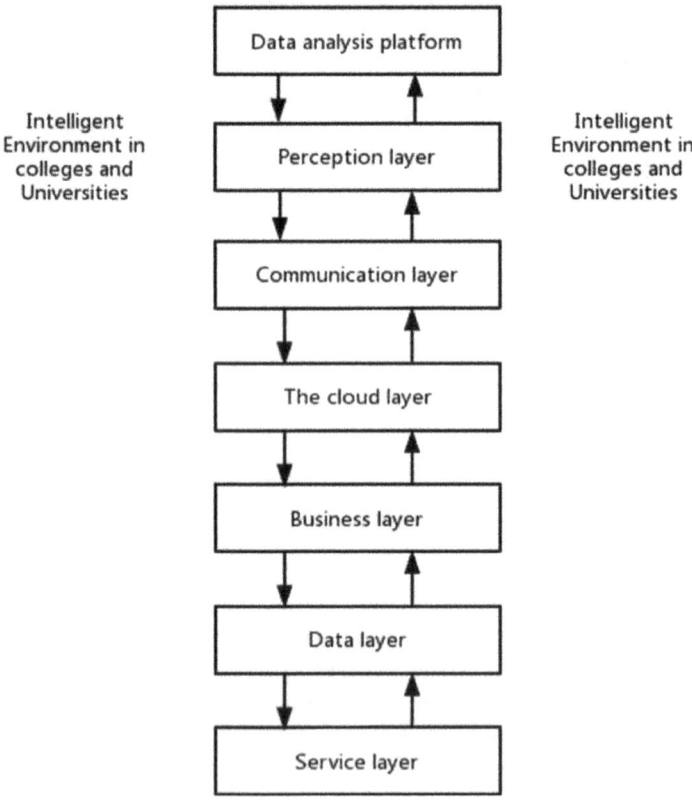

Fig. 1 Smart education system architecture

In the model, the perception layer is responsible for the interaction between the system environment and the external university environment information. Observe the university learning environment and teacher-student information in real time through sensors, mobile terminals and other equipment [4–6], and upload data through the communication layer. The network communication layer is responsible for transmitting the relevant data acquired by the perception layer to the business layer, so as to provide digital network access services for college English language education. Collect the relevant data of students' learning stage and preprocess the data. Pretreatment refers to data cleaning, data regularization, and other processes. After data preprocessing, a data analysis model is built for students' learning behavior data. In this way, it analyzes the learning mode and the degree of mastering knowledge of college students in this semester. Taking the English learning behavior of a college student as the data source, the standard processing process is unified by collecting and sorting the data [7–11].

2   Method

By analyzing the results, we can evaluate the students' mastery of English teaching content and predict the learning effect. Explain the prediction results in detail, establish students' cognitive model, and display the analysis results in all directions by visual technology. The specific steps are as follows: in the evaluation system, according to the students' mastery of English teaching content, the evaluation benchmarks of different evaluation indexes are given, and the corresponding weights of different evaluation indexes are different, so a fuzzy judgment matrix R is constructed, which is used to measure the comparison results of the importance between the evaluation grades and related factors. The matrix form is described as follows:

$$R = \begin{bmatrix} r_{11} & \cdots & r_{1n} \\ \vdots & \ddots & \vdots \\ r_{n1} & \cdots & r_{nn} \end{bmatrix}^* \tag{1}$$

In the formula, r_{nn} represents the membership function. Sum all the factors given in the above formula to obtain $h_1, h_2, \ldots, h_n$, and the weight of factor a_i is

$$\tau = \frac{h_i}{h_1 + h_2 + \cdots + h_n} \tag{2}$$

The above is the method to determine the weight of evaluation index. By comparing experts, the weights of different evaluation indexes are obtained.

All evaluation indicators passed

$$X = \{x_1, x_2, \cdots, x_n\} \tag{3}$$

$$Y = \{y_1, y_2, \cdots, y_m\} \tag{4}$$

Said the evaluation results set, for different evaluation indicators, can be comprehensively described by constructing a fuzzy evaluation matrix:

$$T = \begin{bmatrix} t_{11} & \cdots & t_{1m} \\ \vdots & \ddots & \vdots \\ t_{n1} & \cdots & t_{nm} \end{bmatrix} \tag{5}$$

where t_{nm} represents the corresponding membership function of matrix T. The weight of fuzzy evaluation factor set of the second-level index is given as follows:

$$A = \{a_1, a_2, \cdots, a_n\} \tag{6}$$

Combined with the maximum membership benchmark, we can evaluate the quality of education, predict students' English learning performance in the future, and build a cognitive model.

3   Empirical Analysis

In this paper, students in two classes of a university are selected to use answering machines to participate in classroom teaching, such as signing in, answering questions, taking answers, and rushing to answer questions. Teachers collect and sort out all the behavior data left by students on the learning platform every week. The author did not design and implement teacher intervention measures in the first teaching stage. The platform data details of some students are given in Table 1.

All students participated in the three-stage test, and the senior teachers of the teaching and research group unified the proposition, including sentence translation, single choice, lexical fill-in, sentence pattern conversion, reading and writing, with a total score of 100 points. See Appendix for details. After the test paper is sealed, it is composed of two assistant teachers. The only difference is that some of the

Table 1   Student learning data

Name	Attendance rate (%)	Cumulative response times	Average dispersion velocity (s)	Cumulative scoring rate (%)
A	92	13	14	63
B	100	8	10	58
C	75	3	13	51
D	75	5	17	49

Table 2 Number of test scores for the first time

Grouping number of people	> 80	70–80	60–69	< 60
54	27	9	5	13

students' tests are those who have given the study plan after systematic evaluation, and the other is those who have not touched the information platform. All the data of students' grades will be entered into the "interactive classroom" platform. The grouping details of the test are given in Table 2.

Next, this paper explores the overall influence of information optimization as a shallow intervention on learners' achievement prediction factors after data analysis. Table 2 describes the results of intervention in the second stage of teaching (Table 3).

According to the data in Table 2, from the average changes of the four variables, the average number of answers in class after intervention ($M = 22.16$) is larger than that before intervention ($M = 19.96$), which shows that students can actively participate in the classroom after receiving shallow intervention, which makes the number of answers in class increase; In the number of wrong questions in the current day, the average value after shallow intervention ($M = 6.99$) is larger than that before intervention ($M = 11.51$), which shows that students will pay attention to consolidating the knowledge points they have learned after shallow intervention, so as to reduce the number of wrong questions in the current day; the average answer speed after shallow intervention ($M = 0.79$) is higher than that before intervention ($M = 0.73$), which shows that students will accelerate the average answer speed

Table 3 Comparison of the differences of students' achievement prediction factors before and after information optimization intervention

Variable	Before and after the intervention	M	SD	CI(MD)	T (df)	P (2-tailed)
Number of answers in class	No intervention	19.96	5.88	(−4.31, −0.08)	− 2.13 (26)	0.043
	Shallow intervention	22.16	5.31			
Number of wrong questions of the day	No intervention	11.51	7.75	(0.56,5.77)	2.49 (26)	0.019
	Shallow intervention	6.99	7.81			
Average answer speed	No intervention	0.73	8.13	(1.82,7.22)	3.44 (26)	0.002
	Shallow intervention	0.79	6.49			
Visit transcript	No intervention	19.96	0.32	(− 0.17, − 1.03)	− 1.03 (26)	0.312
	Shallow intervention	24.24	0.25			

after receiving shallow intervention. On the interview transcripts, the average value of interview transcripts after intervention ($M = 24.24$) is larger than that before intervention ($M = 19.96$), which indicates that students can increase the number of interview transcripts and pay attention to their learning situation in time under the guidance and intervention of the system. In addition, the paired sample t-test results show that students have significant differences in three variables: the number of answers in class ($t = -2.13, p < 0.05$), the number of wrong questions in the day ($t = 2.49, p < 0.05$), and the average answer speed ($t = -3.44, p < 0.05$), but in the interview transcript, this shows that shallow teacher intervention can significantly affect the students' performance predictors in three aspects: the number of answers in class, the number of wrong questions in the day, and the average answer speed, but has no significant effect on the visiting transcripts.

4 Conclusion

Under the big data environment, the key to improve the quality of college English language education lies in how to build an open, healthy, and harmonious English language education ecosystem. The application and development of big data technology breaks the balance of various factors in the traditional college English language education ecosystem. Big data technology provides support for building a good ecological environment for college English language education. Through big data technology, we can build a comprehensive, balanced, and reasonable college English language education environment. This paper aims to balance the ecosystem of college English open education by constructing the relevant departments of various educational factors and explore the college English language education mode and method of combining classroom and practice under the big data environment, so as to play a guiding role in improving the quality of English language education and teaching.

References

1. Rist, R.C.: On the application of ethnographic inquiry to education: procedures and possibilities. J. Res. Sci. Teaching **19**(6), 439–450 (2010)
2. Pitura, J., Terlecka-Pacut, E.: Action research on the application of technology assisted urban gaming in language education in a polish upper-secondary school. Comput. Assisted Lang. Learn. **31**(5–8), 734–763 (2018)
3. Ardito, L., Scuotto, V., Del Giudice, M., Petruzzelli, A.M.: A bibliometric analysis of research on big data analytics for business and management. Manag. Dec. **57**(8), 1993–2009 (2019)
4. Wenyuan, C., Ying, Z., Kunyan, L., Rongxian, Z.: Research on public service based on big data:content analysis of the reports of public bicycle in the four places across straits. New Media Soc. 1 (2018)

5. Jinghu, P., Jianbo, L.: Research on spatial pattern of population mobility among cities: a case study of "Tencent Migration" big data in "National Day–Mid-Autumn Festival" vacation. Geog. Res. **38**, 1678–1693 (2019)
6. Ma, D., Hu, J.: Research on collaborative management strategies of closed-loop supply chain under the influence of big-data marketing and reference price effect. Sustainability **12**(4), 1685 (2020)
7. Corwin, E.J., Jones, D.P., Dunlop, A.L.: Symptom science research in the era of big data: leveraging interdisciplinary resources and partners to make it happen. J. Nursing Scholarship **51**(1), 4–8 (2019)
8. Bonacchi, C., Krzyzanska, M.: Digital heritage research re-theorised: ontologies and epistemologies in a world of big data. Int. J. Heritage Stud. **25**(12), 1235–1247 (2019)
9. Pitura, J., Terlecka-Pacut, E.: Action research on the application of technology assisted urban gaming in language education in a Polish upper-secondary school. Comput. Assisted Lang. Learn. **31**(7), 734–763 (2018)
10. Hu, J., Zhang, Y.: Measuring the interdisciplinarity of big data research: a longitudinal study. Online Inf. Rev. **42**(5), 681–696 (2018)
11. Senthil, G., Lehner, T.: Schizophrenia research in the era of team science and big data. Schizophrenia Res. **217**, 13–16 (2020)

CFD Analysis of Droplet Diffusion in Negative Pressure Ward

Liuyang Kong, Qiong Li, Tao Liu, Wenfeng Gao, Mu Zhang, and Yuchong Zhang

Abstract The study of particle diffusion in negative pressure ward can lay a foundation for the application of pollutant diffusion in negative pressure ward based on Internet of things. A negative pressure ward models and droplet models are established, and the effect of droplets with different particle sizes in the negative pressure ward on indoor airflow is simulated. Computational fluid dynamics (CFD) was used to study the movement process, concentration distribution, surface adhesion, and suspension of droplets in the same negative pressure ward environment with particle diameters of 5, 20, 60, and 90 μm. The results show that: when the particle diameter is 5 and 20 μm, the movement and diffusion of the droplets are intense and the droplets mainly enters the exhaust outlet with the flow of air. When the particle diameter is 60 and 90 μm, the influence of gravity on the droplet is greater than that of the airflow in the room, so the diffusion is weak and the droplet particles are more adherent. It is mainly concentrated on the ground, the surface of the bed, and the patients. Weaker discharge effect was obtained when the droplets produced by patients are far away from the exhaust outlet.

1 Introduction

The acute respiratory corona virus (COVID-19) spreads worldwide over a short period of time. The aerosol particles produced by patients from breathing, talking, and coughing in negative pressure wards are regarded as the major source of indoor biological pollution of COVID-19 [1]. By combining the Internet of things with indoor ventilation and connecting sensors with the Internet of things, indoor air distribution parameters can be obtained to provide theoretical support for controlling indoor ventilation, improving the air condition in the negative pressure space, and reducing the probability of droplet infection. Many scholars have studied the particle size distribution of droplets produced by patients, the number of these, the diffusion,

L. Kong · Q. Li (✉) · T. Liu · W. Gao (✉) · M. Zhang · Y. Zhang
Solar Energy Research Institute, Yunnan Normal University, Kunming 650500, China
e-mail: liqiong@ynnu.edu.cn

© The Author(s), under exclusive license to Springer Nature Singapore Pte Ltd. 2021   115
L. C. Jain et al. (eds.), *3D Imaging Technologies—Multidimensional Signal Processing and Deep Learning*, Smart Innovation, Systems and Technologies 236,
https://doi.org/10.1007/978-981-16-3180-1_15

motion of these on different occasions, and the influence of droplets concentration in wards under different gas. Well [2] found that aerosol particles with small particle sizes have a wider diffusion range and are the main propagation medium of aerosol particles in the air. Large particle sizes are mostly sink by gravity, and the propagation of large particle sizes is very different from that of small particle sizes. Yang [3] studied the particle size of droplets produced by cough in healthy people, and the average particle size was 8.35 m, with a three-peak distribution. Liu [4] conducted a simulation study on the influence of different gas tissues on droplet concentration in the common room.

CFD method was used to simulate the movement of different droplets in the emergency negative pressure ward of Thunder God Mountain hospital in Wuhan based on Euler–Lagrange method.

2 The Physical Model

The Thunder God Mountain negative pressure ward was investigated, and the droplet model was configured in Fluent in combination with the droplet particle model (DPM). Assuming that the mouth is a surface jet source, the droplet pollutants exhaled by human body from the mouth are a mixture of continuous phase (air) and discrete phase (droplet particles), and the spray velocity of droplet and air is approximately the same (Table 1).

2.1 Physical Model of the Ward

Combined with the actual situation of the standard infectious ward of Thunder God Mountain negative pressure, we can get it. The flow rate of air intake is 500 m³/h while that at air outtake is 700 m³/h. See Fig. 1 for detailed dimensions.

Table 1 Configuration of environment

Parameter	Value	Parameter	Value
Velocity at Inlet1	3.93 m/s	Infectious ward	6 m * 3 m * 2.6 m
Velocity at Inlet2	3.93 m/s	Bed	2 m * 1.2 m*0.6 m
Velocity at Inlet3 (patient's mouth 1)	$y = 0.3\sin\left(\frac{\pi}{2} * t\right)$	Distance between outlet1 and bed1	0.2 m
Velocity at Inlet4 (patient's mouth 2)	$y = 0.3\sin\left(\frac{\pi}{2} * t\right)$	Distance between outlet2 and bed2	2 m
Differential pressure in the room	− 5 Pa	Radius of the air inlet	0.15 m

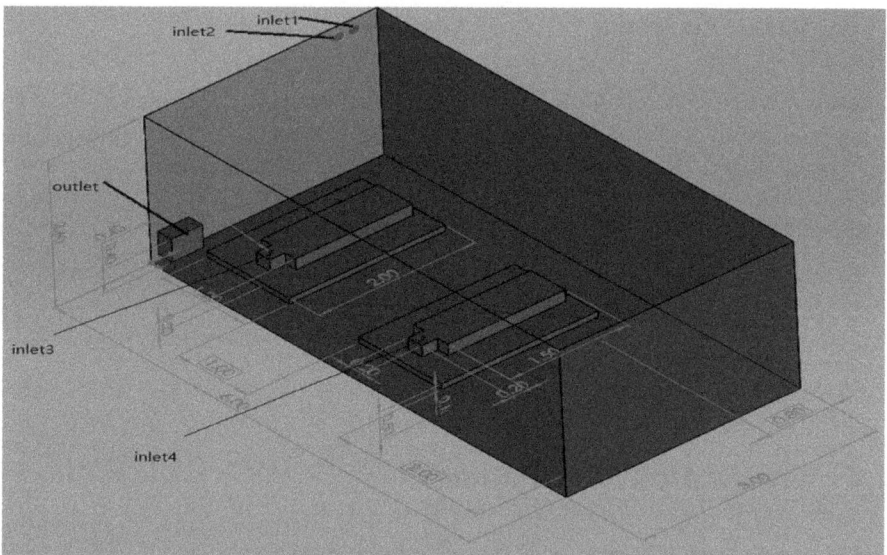

Fig. 1 Diagram of negative pressure ward

2.2 Droplet Model

It is found that the particle size and distribution of droplets obtained by different test methods are very different. The National Health Commission [1] pointed out that the transmission characteristics of COVID-19 droplets are generally larger than 5 μm in particle size, and droplets between 4 and 20 μm in the air carry the most viruses. In this paper, Liu [4] classified and analyzed the experimental data of different literatures and analyzed the analysis results of droplets produced by human speaking and breathing activities (with particle size range of 0–200 um), and combined with the transmission characteristics of COVID-19 droplets. In this paper, 500 droplets are generated per second, and the particle size of simulated droplets is configured as follows:

(1)   There are four kinds of droplet sizes: 5, 20, 60, and 90 μm.
(2)   Percentage of droplet size produced by breathing and speaking: 5 μm (15%), 20 μm (9%), 60 μm (48%), and 90 μm (28%).
(3)   The continuous phase calculation is realized based on Euler method, the discrete phase calculation is realized based on Lagrangian method, and the influence of gas flow on particle diffusion is considered. The droplet particles are smooth solid spheres and remain spherical throughout the movement.
(4)   Layout of droplet source: The droplet source on the bed surface is the circular r = 0.02 m of the mouth of the patient. The droplet particle density is 1003 kg/m^3, and the patient's normal exhumation velocity of droplet is 0.3 m/s.

2.3 Breathing Model

Since the injection velocity of different respiratory activities of human body is not a constant value, such as the injection velocity of respiration is a sine function and the injection velocity of cough is an impulse function. Therefore, in this paper, user-defined function (UDF) is adopted to solve the problem. The respiratory formula is obtained through Gupta JK's research on respiratory conditions [5], and the formula is:

$$y = 0.3 \sin\left(\frac{\pi}{2} * t\right) \tag{1}$$

2.4 Physical Model Simplification

The above physical model is simplified as follows: Heat exchange is not considered. According to the room size and air intake, the Re number is more than 105 and determined as a turbulence model to calculate the indoor flow field. The shape change and material change of droplet size are not considered. Magnus forces, electric forces, magnetic forces, and thermal and electrophoresis phenomena on the particles are not considered. The dust source is not considered to be subjected to Saffman force in the adjacent wall. The collision between the droplets and the wall surface is set as a rebound. Droplets are attached to the surface of the bed and the patient's surface is set for complete adhesion.

3 The Governing Equation

3.1 Governing Equation of k-ε Model in Gas Phase

The k-ε two-equation turbulence model is presented, and the numerical solution is the turbulence model. The two-phase flow equation of intra-room turbulence is as follows:

$$div(pv\phi) = div(\Gamma \operatorname{grand}\phi) + S_\phi + S_p \tag{2}$$

In Eq. (2), ϕ is the independent variable, Γ is the generalized diffusion coefficient, S_ϕ is the original term of the gas phase, and S_p is the original term of the particle.

3.2   Particle Phase Control Equation:

The DPM model is used in this paper, in the Lagrangian coordinate system, for example, the momentum equation of the droplet particle considering only the resistance and gravity of the particle, is as follows:

$$\frac{dv_{ki}}{d_t} = (v_i + v_{ki})/\tau_{rk} + g_i + g_s \tag{3}$$

In Eq. (3) V_i is gas phase velocity, particle phase velocity in the direction of i, τ_{rk} is particle dynamic relaxation time, g_i is gravity acceleration, and g_s is Saffman force.

4   Numerical Simulation Results and Analysis

4.1   Validation

Figure 2 shows the indoor air flow obtained by numerical simulation. The air flow in the air inlet is blocked by the wall after flowing, resulting in circumfluence, which is attached to the wall flow. Finally, the air flow in the air inlet is captured by the air flow in the exhaust port after passing through the hospital bed. Such gas structure can greatly discharge the air pollutants produced by patients.

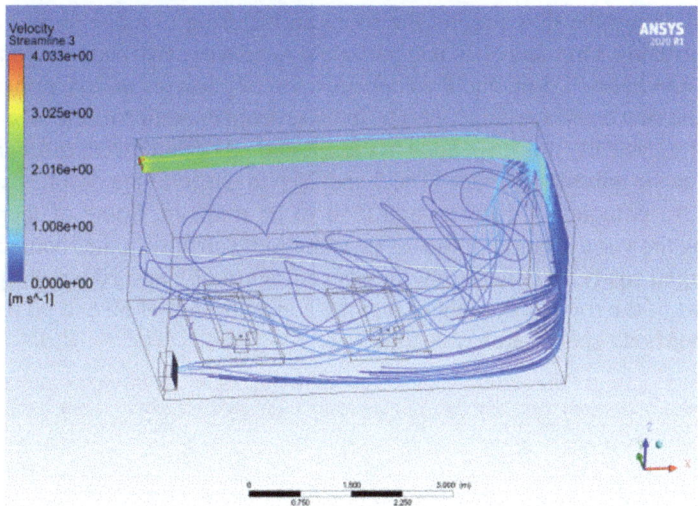

Fig. 2 Indoor air movement

The gas flow in the negative pressure ward obtained by simulation in this paper is consistent with the gas structure law obtained by Zhang [6] in the ventilation system design of the negative pressure ward in Thunder God Mountain hospital. It is shown that the modeling accords with the actual gas flow in the negative pressure ward.

4.2 Analysis of Concentration Distribution of Droplets with Different Particle Sizes

Figure 3 shows the gradual diffusion of droplets with different particle sizes generated by the patient's breathing and speaking in the negative pressure ward along with the airflow movement within a patient's respiratory cycle of 4 s. According to the images within 30 s, 5 and 10 um particle sizes moved toward the exhaust outlet and the middle of the room, respectively, under the influence of room airflow. Different particle sizes began to be divided into groups according to particle size due to different velocities, but they were not fused and diffused with the indoor airflow, and the droplet concentration was relatively high. The 60 and 90 μm particle size droplets generated by the patient away from the exhaust in 30 s were less affected by indoor fluency and more affected by gravity, and began to scatter in the front half of the bed within 10 s, resulting in higher concentrations of 60 and 90 μm particle size droplets in the bed body and the upper half of the patient. Droplets with particle sizes of 60 μm and 90 μm produced by patients close to the exhaust outlet did not diffuse within 30 s and moved toward the exhaust outlet with the influence of the exhaust airflow. Concentrations of particle sizes of 60 and 90 μm were higher on the path connecting the patient's head to the exhaust outlet. The images in 60 s showed droplets of 5 and 20 μm following the flow around the room and starting to move around the room without diffusing. Droplets of 20 μm particle size between two beds are concentrated in suspension below 1.5 m, and the concentration of particles of 20 μm particle size between the two beds is relatively high. In 60 s, droplets with particle size of 60 and 90 μm were basically not suspended in the air, and all of them were discharged from the room or the adhesion surface. The 5 and 20 μm particle size droplets suspended in 1990s are completely diffused and tend to be uniform. Some of them will be affected by the vortex region of the upper and middle airflow in the room and cannot be discharged easily. In conclusion, the concentration of 5 and 20 μm droplet in the upper heart of the room was highest, and the concentration of 60 and 90 μm droplet in the upper body spatial area was highest.

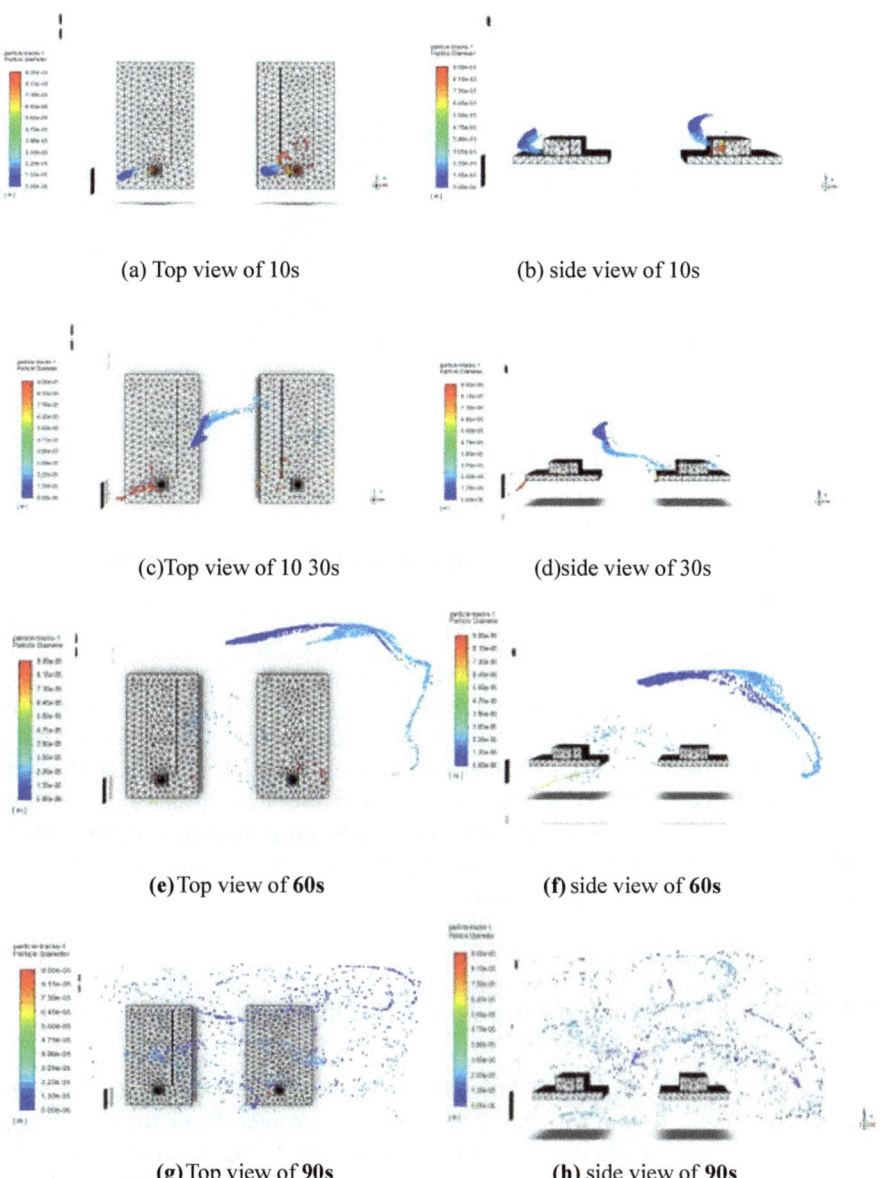

(a) Top view of 10s　　　　　　　(b) side view of 10s

(c)Top view of 10 30s　　　　　　(d)side view of 30s

(e) Top view of **60s**　　　　　　(f) side view of **60s**

(g) Top view of **90s**　　　　　　(h) side view of **90s**

Fig. 3 Dispersion of droplets at different times

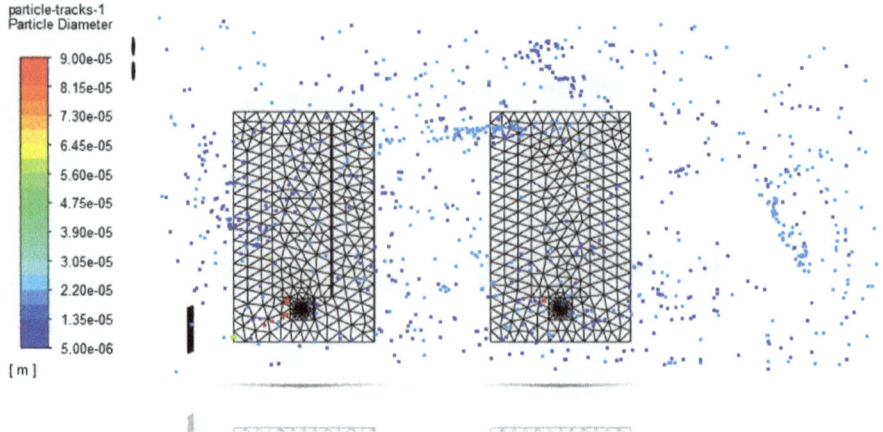

Fig. 4 Particle size distribution map of 150 s

4.3 Adhesion Locations of Droplets with Different Particle Sizes

After 150 s, the droplets in the room tend to be uniform. When the droplets of different particle sizes produced by the patient tend to be evenly distributed in the negative pressure ward, the adhesion between the bed and the patient surface is analyzed in Fig. 4. The distribution of 60 and 90 μm droplets are affected by the distance between the bed and the outlet. For the bed close to the outlet, droplets are distributed near the patient's head. For the other bed, droplets are distributed around the bed. Droplet size of 20 and 5 μm is evenly distributed, basically floating in the air, and the phenomenon of concentrated adhesion is not obvious.

4.4 Discharge Effect of Droplets with Different Particle Sizes

According to the analysis of the relationship between the discharge quality and time of the exhaust outlet within 160 s, as shown in Fig. 5. Droplets begin to be discharged from the 12 s vent, and the discharge mass is less than 25 mg between 12 and 25 s. Combining with the concentration distribution of droplets at 10 and 30 s in Fig. 3, it can be concluded that all droplets discharged during this period are 5 and 20 μm particle size droplets produced by patients near the vent. Between 25 and 40 s, the mass of the exhaust outlet increased sharply to 450 mg, which was obtained by combining the images in Fig. 3, 30 and 60 s. Droplets 60 and 90 μm in the proximal vent and 5 and 20 μm in the distal vent at this time enter the vent and increase the discharge mass. The mass flow of the outlet after 60 s decreased significantly and tended to be flat. The mass flow from 90 to 110 s increased from close to 0 m–30 mg.

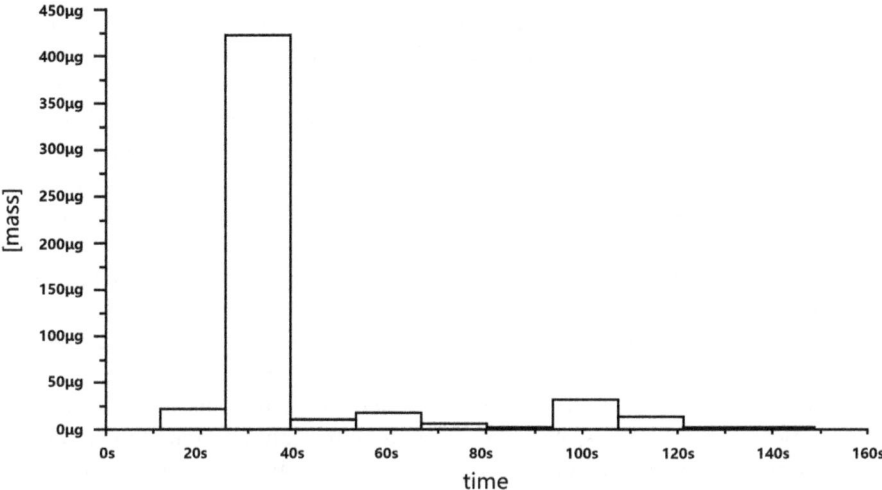

Fig. 5 Quality of droplets discharged from the vent at different times

Based on the analysis of Fig. 3, it was found that the 60 μm particle size droplets produced by the patient at the far exhaust outlet and the 5 and 20 μm particle size droplets discharged from the far exhaust outlet were captured by the exhaust airflow after circulating around the room, resulting in the increase of the mass flow at the exhaust outlet. It can be concluded that in the case of air distribution in the negative pressure ward, the droplets produced by patients close to the exhaust outlet can be discharged within 40 s. The efflux time of the droplet produced by the patient away from the vent is long, and the effect is poor.

5 Conclusion

The results of the data simulation study show that:

(1) The diffusion of small volume droplets represented by droplets with a particle size of 5 μm is basically similar to the airflow flow. In a short time, small particle size droplets are easy to be suspended in the space, and the density is high around the wall and in the vortex area in the middle of the room. The volume mass of the droplet with a particle size of 20 μm is slightly larger than that with a particle size of 5 μm. Basically, the droplet will not be suspended in the self-locking region, but will float around the room following the air inlet. In the negative pressure ward, 90 and 60 μm particle size droplets were near the air outlet ward and were not affected by the air outlet for a short time, and the concentration was the highest near the patient's head.

(2) Large particle size droplets produced by the patient at the distal end of the exhaust outlet were mostly deposited near the hospital bed at the distal end of

the exhaust outlet and sank to be captured by the hospital bed and the patient surface. At the end of the inlet and outlet, large particle size droplets are mostly concentrated the head, and adhere to the external surface. In addition, the droplet nuclei with large particle size are relatively large, so it should be noted that the droplet nuclei should be far away from the surface of the patient with the exhaust outlet and the surface of the patient's head close to the air outlet. The concentration of droplet droplets and the concentration of adhered virus is high, which can enhance disinfection.

(3) The discharge effect of large particle size droplets (90, 60 μm) generated by the far exhaust outlet in a short time is poor. Due to the poor float ability of large particle size droplets, the discharge effect is significantly better than that of small particle size droplets. Because the small particle size is easy to be suspended, the small particle size generated by patients in the far exhaust outlet is easy to follow the airflow to float to the spiral region in the middle of the room, which is not conducive to the exhaust outlet discharge. The patient and the vent affect particle delivery until the droplet has been fully diffused.

References

1. The National Health Commission. Covid-19 diagnosis and treatment protocol (6 edn) (2019)
2. Well, F.W.: On air-borne infection study II droplets and droplet nuclei. Am. J. Epidemiol. **20**, 611–618 (1934)
3. Yang, S., Lee G.W.M., Chen C.m., et al:. The size and concentration of droplets generated by coughing in human subjects. J. Aerosol Med. Official J. Int. Soc. Aerosols Med. **4**, 484–494 (2007)
4. Liu, P.: Concentration Variation Characteristics and Pollution Control of Droplet Pollutants in Wards. Chongqing University (2016)
5. Gupta, J.K., Lin, C., Chen, Q., et al..: Characterizing exhaled airflow from breathing and talking. Indoor Air **1**, 9–31 (2010)
6. Zhang, S., Cheng, M., Gao, W., et al.: The application of CFD technique in the design of ventilation system of negative pressure ward in Leishenshan Hospital. Huazhong Archit. **38**(4), 78–81 (2020)

Research on Cognitive Analysis Method of Command Mechanism Based on Operational Data Flow Motifs

Chen Yuan, Bo Du, and Xiangwei Xiao

Abstract The key to the efficient operation of combat system is combat command, and the key to the efficiency of combat command is command mechanism. Aiming at the cognitive problem of command mechanism in combat system, this paper constructs a multidimensional view model of combat system, clarifies the internal relationship among operational action, operational data, and command mechanism from the perspective of cross-domain analysis. Proposes a command mechanism analysis method based on operational data flow motifs (ODFM), gives the definition and graphic representation of operational data flow motifs, and deeply discusses the action mechanism of command mechanism. Finally, around the three typical command mechanisms, this paper combs the key elements of the mechanism construction, gives the basic ideas of the command mechanism construction, and provides ideas for the scientific cognition and construction of the command mechanism.

1 Introduction

The current form of war has entered the era of network information system warfare, which is dominated by information, centered on network, and based on data. Battle command is the key factor to affect the operation efficiency of combat system, and the key of battle command running efficiency depends on command mechanism. As artificial intelligence, edge computing, cloud computing, big data, such as the rapid development of technology, especially the big data in innovation in the field of military application, is an unprecedented activate data, releases the data value, has a multiplier effect, speeds up the reshaping command process, acting as a whole, combat big data is becoming the key factor of influence and change command mechanism.

Combat, as an important form of non-war military operations, involves many elements such as command decision, force delivery, disposal, and comprehensive

C. Yuan · B. Du (✉) · X. Xiao
Engineering University of PAP, Urumqi 830049, China

support, the interaction among entity elements is complex, and it is a typical systematic combat. By exploring the formation mechanism of command mechanism under the condition of big data, there is a growing demand to reveal the rules of systematic combat and study the way to victory. However, the traditional research methods is given priority to summary of experience and operational practice, under the background of the operational pattern changed to informatization, combat space to the domain extension, operational factors to diversified evolution, the combat data is exponentially increasing, command mechanism study the following two challenges: one is a component unit of operational system which has complex interactive relationship, and architecture is in the process of dynamic change in real time. With the characteristics of self-adaptation, self-organization, unclear causality, and non-unique combat effect [1, 2], how to sort out, depict, and explain the formation mechanism of command mechanism in this complex interaction; secondly, the system operations in essence are a kind of distributed cross-domain operations [3], involving physical domain, information domain, and cognitive domain multiple fields such as space, only from the cognitive domain, namely "commander" perspective brought to the analysis of forming regularity consideration command system, it is impossible to get overall conclusions, how to in the process of research from the perspective of multidimensional understanding command mechanism, operational data, the relationship between operations.

In the face of the above problems, it is obviously difficult to understand and reveal the rules of mechanism formation from the perspective of "macro" cognition. Starting from the perspective of complex scientific research, this paper first proposes a multidimensional view model of the combat system from the perspective of systems engineering methodology, and clarifies the internal relationship between combat operations, combat data and command mechanism. Then, a command mechanism analysis method based on Data Flow Motif (DFM) was proposed, and the mechanism of command mechanism was analyzed. Finally, the basic idea of command mechanism construction is put forward.

2 Multidimensional View Model of Combat System

Referring to the theory of multidimensional view proposed by Zachman [4], multidimensional view is essentially a methodology for system design. It analyzes the same complex problem from the perspective of people in different fields and forms several interrelated views, so as to comprehensively and accurately describe the overall architecture of the system. Considering the distributed cross-domain characteristics of the combat architecture, this paper selects three perspectives of combat data, combat operations, and command mechanism to build a multidimensional view model of the combat system, as shown in Fig. 1. Through retrospective analysis, the internal relationship between operational data, command mechanisms, and operations can be explained.

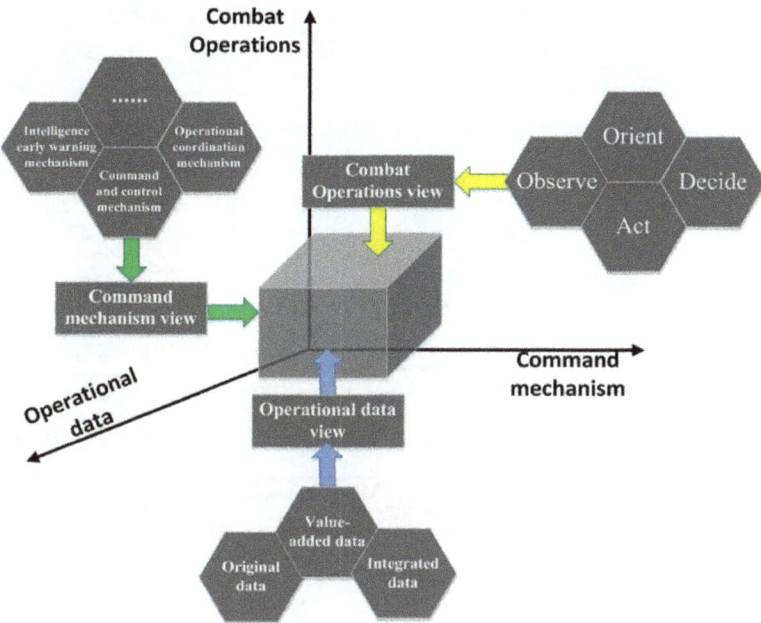

Fig. 1 Multidimensional view model of combat system

2.1 Combat Data View

The combat data view is an abstraction of the combat system from the perspective of the combat data. It is the dynamic change of various attributes of the combat entity described, recorded, and represented by the data stream in the process of system confrontation. Starting from the data value density, it can be divided into three categories: original data, integrated data, and value-added data. Among them, raw data refers to the unprocessed data collected and collected through private network, special line and public network. Integrated data is a data resource that is cleaned and associated with the original data according to the unified classification, transformation, governance, and other data standards and norms of our situation, enemy situation, and combatfield environment. Value-added data is a subject-oriented data product that is formed to support operational applications through data fusion and consolidation on the basis of integrated data, such as comprehensive situation and knowledge map of operational data. The operational big data resource architecture is shown in Fig. 2.

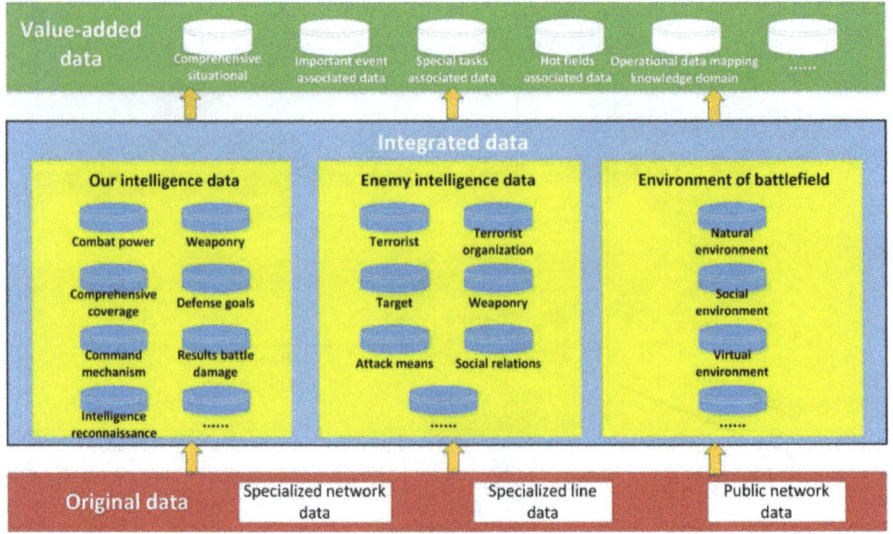

Fig. 2 Operational big data resource system framework

2.2 Command Mechanism View

Command mechanism view is the abstraction of combat system from the perspective of command mechanism. Command mechanism refers to the operation rules, processes, and modes of combat command system. The connotation of command mechanism varies with different tasks. For example, when dealing with large-scale group incidents or public security emergencies, it generally includes intelligence early warning mechanism, emergency response mechanism, decision-making and planning mechanism, control coordination mechanism, joint cooperation mechanism and comprehensive guarantee mechanism, etc. Generally speaking, according to the operation season, the command mechanism generally includes intelligence and early warning mechanism, command and control mechanism, operation cooperation mechanism, etc.

2.3 Operational View

Combat operations view is an abstraction of combat systems from the perspective of combat operations. In the description of the combat process, Boyd, an American military theorist, put forward the OODA (Observe-Orient-Decide-Act) ring theory in the 1970s. Under the condition of information, the combat, as a typical systematic combat, can be regarded as a systematic OODA ring process [5]. Thus, the combat

Table 1 Relationship description matrix

Perspective	Elements				
	Domain space	Behavior main body	Specific operation	Constituent elements	Multidimensional view
Operations	The physical domain	Commanders and fighters	Reconnaissance, decision making, strike, etc	Intelligence and reconnaissance network, weapons platform network, material support network	Operational view
Operational data	Information domain	Operational data support personnel and technical personnel	Collection, fusion, analysis, application, etc.	Sensor network, data fusion processing unit, infrastructure	Combat data view
Command mechanism	Understanding the domain	The commander	Constrain, standardize, judge and optimize the process	Intelligence early warning mechanism, command and control mechanism, operational coordination mechanism	Command mechanism view

process can be abstractly described as the iterative cycle of four kinds of operations: "Observe"–"Orient"–"Decide"–"Act."

Thus, it can be seen that combat data, command mechanism, and combat operations are essentially abstractions of combat systems in different dimensions, and there is an interactive relationship between the three views. Table 1 sorts out the relationships among domain space (Where), behavior subject (Who), specific operation (How), constituent element (What), and multidimensional view (View).

3 Analysis Method of Command Mechanism Based on ODFM

3.1 Basic Ideas and Related Concepts

Combat system is essentially a distributed complex network, involving multiple domains such as physical domain, information domain, and cognitive domain. It can be abstractly abstracted into a hypernet structure constructed by physical network, information network, and command network [2]. The emergence of system capabilities is actually generated through the coupling relationship between component systems and the complex information interaction. The command mechanism is the operating rules, procedures, and methods of the command system, which belongs to the cognitive domain. By referring to the super-network theory [6], the command mechanism can be abstracted from the nodes in the hypernetwork structure, and its mechanism can be realized by analyzing the interaction between this node and other nodes (sensor nodes, charge nodes, combat nodes, etc.).

According to the concept of network motifs proposed by Milo et al. [7], the interaction between nodes can be characterized by operational data flow motif (ODFM), and the operational data flow and operational data flow motifs are defined as follows:

Definition 1: Combat data flow refers to a complete data function link that generates, processes, and uses data in the combat architecture.

Definition 2: The operational data flow motifs refers to the data action mode that repeatedly occurs in the operational architecture to produce, process, and use data so as to realize certain system functions.

3.2 Construction of Operational Data Flow Motifs

Formal description of hypernetwork nodes. According to the above definition, the information age combat model (IACM) [9] was proposed by referring to the classification method [8] and Cares. The subsystems of the combat system were abstracted into five types of nodes, namely sensor nodes S, decision-making nodes D, Combat nodes I, target nodes T, and command mechanism nodes M. The node definition is as follows:

Definition 3: Sensor nodes S represent the entities of reconnaissance, detection, and monitoring, and have the capability of dynamic perception of combatfield space. The main function is to observe, identify, and probe the combat targets, and transmit the acquired target information to the decision nodes in the form of information flow. Sensor nodes such as infrared radar, UAV, and reconnaissance satellite. It can be formally expressed as a collection of sensor nodes.

$$S = \{S_1, S_2, \ldots, S_n\} \tag{1}$$

Definition 4: The decision-making nodes D represent the command, control, and decision nodes, which have the abilities of command, control, intelligence analysis, processing, fusion, and judgment. Its function is to integrate and identify the target information of the sensor node or other decision-making nodes, make decisions, and command the influencing nodes to carry out combat operations. It can be expressed formally as a set of decision nodes.

$$D = \{D_1, D_2, \ldots, D_m\} \tag{2}$$

Definition 5: Operational nodes I refer to the operational entities of sealing and control, encirclement and suppression, pursuit and destruction, setting up a card and setting up an ambush, which have the capabilities of blockade, control, attack, pursuit, interception, and annihilation. Its function is to receive the instruction from the decision-making node, deploy, and carry out combat operations against the target, such as sealing and control operations, encirclement and suppression operations, and pursuit and annihilation operations. It can be formally expressed as a collection of operational nodes.

$$I = \{I_1, I_2, \ldots, I_p\} \tag{3}$$

Definition 6: Target nodes T represent all kinds of enemy nodes, including terrorist leaders, terrorist organizations' camps, terrorists who carry out terrorist attacks, and other operational targets related to terrorist organizations. It can be expressed formally as a collection of target nodes.

$$T = \{T_1, T_2, \ldots, T_q\} \tag{4}$$

Definition 7: The nodes M of command mechanism represent various command mechanisms, generally including intelligence early warning mechanism, command and control mechanism, operational coordination mechanism, etc. The command mechanism varies according to the different tasks undertaken by the system. It can be expressed formally as a collection of mechanism nodes.

$$M = \{M_1, M_2, \ldots, M_r\} \tag{5}$$

Definition of operational data flow motifs and graphical representation. Based on the above node definitions and flow motifs definitions, this paper proposes three types of data flow motifs for operations, whose classification, graphical representation, and sample description are given in Table 2. The specific definition is as follows:

Definition 8: Information data flow motif (IDFM) refers to an Information functional link for the generation, processing and use of intelligence Data.

Definition 9: Command and control data flow motif (C2DFM) refers to an information functional link for the generation, processing and use of charge Data.

Definition 10: Cooperative data flow motif (CoDFM) refers to an information functional link for the generation, processing, and use of state data.

Mechanism analysis based on DFM command mechanism. The command mechanism is analyzed based on three basic data flow motifs, which can reflect the function and action mode of the command mechanism in a certain combat link of the combat system.

The mechanism of intelligence early warning mechanism can be explained by analyzing the information data flow motifs. Taking target detection as an example, after the sensor node S in the physical domain perceiving and acquiring the target data, it can distinguish through the intelligence warning mechanism M_1 constructed in the

Table 2 Basic data flow motifs

Data flow motifs type	Basic data flow motifs representation	Instance to describe
Information data flow motifs (IDFM)	$S \rightarrow M_1 \rightarrow D,\ I$	The sensor node will transmit the intelligence data to the decision node or to the combat node after judging by the intelligence warning mechanism
	$T \rightarrow M_1 \rightarrow S'$	The sensor node determines which combination of nodes is used to detect the target node through the intelligence warning mechanism
Command & control data flow motif (C2DFM)	$D \rightarrow M_2 \rightarrow S^+,\ I^+$	The decision-making node sends detection instructions to the sensor node step by step through the command and control mechanism. Or send an attack command to the combat node
	$I^+ \rightarrow M_2 \rightarrow T$	The operational nodes use the command and control mechanism to determine which attack nodes will attack the target nodes
Cooperative data flow motif (CoDFM)	$D \rightarrow M_3 \rightarrow D^+$	The cooperative function of decision nodes can be realized by cooperating with other decision nodes through the combat cooperative mechanism
	$S \rightarrow M_3 \rightarrow S'$	The sensor nodes cooperate with other sensor nodes to realize the target detection function through the combat system mechanism
	$I \rightarrow M_3 \rightarrow I^+$	The combat node collaborates with other combat nodes to achieve the target attack function through the combat coordination mechanism

Note + represents 1 or more node units

cognitive domain. If the decision unit is required to further identify and judge the target data, it can transmit the relevant target data to the decision node D. If the information accuracy, information delay and operational requirements meet the requirements of firepower, the target detection data will be transmitted to the operational nodes I.

The function mechanism of command and control mechanism can be explained by analyzing command and control data flow motifs. Taking operational decision as an example, the decision-making node D needs to judge through the command and control mechanism M_2 after the comprehensive analysis of the intelligence information is obtained. If the current intelligence data accuracy is not enough to support the attack on the target, then send a continuing detection command to the sensor node S. If the strike conditions are met, the appropriate combat node I is directly selected to issue the command and strike the target.

By analyzing cooperative data Flow motifs, the mechanism of operational coordination mechanism can be explained. Decision together, for example, because detecting node controlled by each decision node D is different, the collection of the target data is different, and between decision node belongs, support and security relationship, the attachment is relatively complex, so the cooperation between decision node need to achieve by operational synergy mechanism M_3, in order to achieve the comprehensive decision making, optimize coordination, and ultimately achieve the purpose of the adaptive coordinated. Similar sensor nodes can realize the in-depth perception, recognition, and tracking of the target through the cooperative mechanism, and combat nodes can realize the coordinated attack on the target through the cooperative mechanism.

Through the analysis and description of the functional mechanism of various command mechanisms, theoretical guidance is provided for the construction of sound and perfect mechanism. It should be pointed out that a single data flow motifs reflects the capability and data action mode of a certain combat link or component unit, and cannot describe the overall capability of the combat system. Therefore, through the combination of the basic data flow motifs, the operation mode and operation mechanism of the command mechanism in the system operation process can be described more clearly. See Table 3.

4 Basic Assumption of the Construction of Command Mechanism

Based on the above analysis, the research believes that the construction and improvement of the command mechanism should be based on the current national public emergency response mechanism to ensure the reality and universality. In addition, it should be integrated with the current command system, fully consider the characteristics of carrying out tasks, and highlight pertinence and actual combat. We should also draw on and apply the new generation of information technologies, such as

Table 3 Combined data flow motifs

Data flow motifs assembly	Data flow motifs representation	Instance to describe
IFlow + C2Flow		Sensor to target strike link (integrated strike ring); Sensor detection link based on situation (detection ring)
IFlow + FFlow		Situation sharing link (decision ring); Damage synergy (execution ring)
C2Flow + FFlow		Task-based sensor cooperative detection link (detection ring); Response operation node control link (control ring)

Note: + represents 1 or more node units

artificial intelligence, edge computing, and big data, to energize the combat system and upgrade its capabilities.

4.1 Intelligence Warning Mechanism Should Highlight "Intelligent Cognition"

Situational awareness is the core element of charge control and the premise and basis of effective decision making and correct action. Every operational decision process must start with situational awareness and take its results as the basis of command decision. As an important link to carry on situational awareness and command decision, intelligence early warning mechanism should be built to improve the level of intelligent cognition of the situation.

First, in the aspect of situational awareness, it focuses on the deep integration and efficient application of massive multi-source heterogeneous combat data, so as to enable the combat system to have more profound combatfield situational awareness, so as to realize the autonomy and intelligence of combatfield situational awareness. It should be emphasized that this situational awareness is cross-modal and cross-domain in terms of interaction, and can generate large-scale cascading effects through coupling.

Two is on the situational understanding, using machine learning and artificial intelligence technology, by building a situational cognitive model, simulate the commander situational cognitive process, make computer semantic understanding ability, so as to capture the commander of the subjective consciousness and decision-making experience and as a cognitive product output, to solve the big data under the condition of information overload on cognitive burden of the commander, on the other hand also can break through commander cognitive limitations, make up the shortfall, correcting errors, enhance the commander's cognitive ability, auxiliary commander (people) get faster, more complete, more accurate, and deeper cognitive outcomes.

The third is from situation sharing. In 2017, the Strategic Technology Office (STO) under DARPA proposed the concept of "Mosaic warfare" and thus defined the concept of intelligent tactical edge. The basis of intelligent tactical edge implementation is the security joint sharing of distributed combat data, while the artificial intelligence algorithm runs in multiple scattered edge nodes, and the frequent communication between nodes is faced with the threat of multiple network security attacks. Therefore, trusted local management of situational awareness units is carried out in tactical edge cluster, and lightweight alliance blockchain consensus algorithm and container choreography intelligent contract are designed, so as to realize secure data fusion and sharing.

4.2 Command and Control Mechanism Should Highlight "Intelligent Control"

Command and control is the "glue" of operational elements, the "multiplier" of operational effectiveness and the "nerve center" of operational command. Command and control mechanism is to ensure the efficient operation of the combat system, the key link to ensure smooth access to combat instructions, and also an important part of the command system construction.

First, from the functional perspective, the rationality of the design is constructed according to the principle of "flexible reorganization, coordinated application, and on-demand service," so as to meet the requirements of real-time information interaction, accurate command and control, and reasonable resource allocation. The flexibility of the guarantee can not only support the vertical to the end of the tree command relationship, but also meet the level of intensive, horizontal to the edge, can be dynamically raised enough flat command relationship, in order to adapt to the changes in the command process and establishment system; the robustness of the structure ensures that the mechanism can still implement efficient command and control over the combat system and continue to ensure the completion of the mission when the operational nodes are in effect or performance declines.

Two from the perspective of decision, command, and control is the premise of the commander of the current situation of scientific judgment and correct decisions, on the one hand, can be used for the depth of the incomplete information to enhance learning technology [10], to the depth of the massive situational data correlation analysis, explore potential value of operational data, implementation in the information, data and resources incomplete scenarios operational situation of identification, analysis and response. On the other hand, through the construction of a decision-making loop based on the "man–machine fusion" mode, the director's qualitative, judgment, and decision-making behaviors can be deeply integrated with the quantitative, analytical, and learning behaviors of the machine at the decision-making level. It should be pointed out that this "fusion" is not a simple "superposition," still less a "combination" based on the combination of rules, but a unified decision-making body based on the sharing, division of labor, interaction, and system integration at the cognitive level.

Third, from the perspective of operational data, command and control essentially relies on the efficient use and rapid flow of operational data, thus shortening OODA time and improving operational effectiveness. On the one hand, based on the concept of "network-centric," we build an integrated adaptive network of "cloud-side-end," which supports the command and control interaction of flexible and dynamic control. On the other hand, based on the study frame, the research of distributed artificial intelligence model of collaborative training, reasoning, and the migration method of study, make full use of the tactics of many distributed on the edge of the combat unit of computing resources, collaborative applications for artificial intelligence model, calculate storage capacity is limited, in combat unit under the condition of network conditions demanding, meet the demand of controller in the real-time data analysis, intelligent decision.

4.3   The Combat Coordination Mechanism Should Highlight "Intelligent Coordination"

The operational coordination mechanism is the core element to ensure the cooperative operation of all operational units. The traditional methods such as planning coordination, contingency coordination, and task coordination cannot meet the requirements of the information age. With the advancement of combat data engineering and the continuous improvement of combat data service system, it will become the general trend to build an intelligent scheduling cooperative mode based on deep data fusion. The key to realize intelligent cooperative scheduling is.

One is to explore both decision-making units, "the cloud-side end" of the combat unit integrated joint operations coordination pattern, form is given priority to with the "cloud" end the brain, to expand and strengthen the "cloud" intelligence "edge" and "end" sinking efficient coordination model, to ensure that the data of dynamic who is "fresh," and build "brain" and the "nerve endings" dynamic loop, achieve organizational coordination flexible, elastic demand assignment of resources, extensive in access, quick access to information and autonomic computing "cloud edges" integrated joint operations coordination model.

Second, in terms of operational data security resource scheduling, use of depth of reinforcement learning techniques, such as design "cloud-edge–end" method of global scheduling optimization near the task "and" the resource cache strategies (such as deployment at the side end are in urgent need of geographic information data, to improve operational support capability based on geographic information data of real time and reduce the invalid information transmission path hierarchy), optimizing calculation uninstall strategy of "the end" and multiterminal coordinated ability (such as artificial intelligence model was used to optimize task executes the main body, and strengthen the organization coordination). Then, the barriers of multi-channel, multi-directional, and multi-source data fusion can be broken through to discover the potential value of combat data, so as to provide dynamic data adaptation and flexible expansion capabilities and realize data-level fusion, feature-level fusion, and decision-level fusion, so as to provide a reasonable and effective basis for collaborative scheduling in real time.

5   Conclusion

This paper analyzes the network information system operating time, the issues, and challenges facing the battle command mechanism, aiming at command in the system of operation mechanism of cognitive problems, put forward the operational system of multidimensional view model, combed the operational data resources system, from building the relationship between the combat system description matrix, the perspective of cross-domain analysis to space (Where) from the field, the behavior main body (Who), a specific operation (How), components (What), multidimensional view (View) five dimensions to clarify operations, operational data, and command mechanism of internal relations. To combat the systematism, networking, distributed and across performance characteristics, puts forward a kind of command mechanism analysis method based on the data flow model body, the concept of three kinds of basic data flow die bodies are defined, and based on the data flow model, body mode and operation mechanism of the function of the command mechanism are analyzed and provided ideas for command mechanism of cognitive science. Finally, based on the cognitive analysis, the paper puts forward the basic idea of building the combat mechanism.

Under the condition of informatization, the modern combat fraught with uncertainty, the emerging of the system capacity, operational effectiveness of high and low is closely connected with command mechanism run efficiently, traditional cognition, and analysis method based on "experience" and research paradigm, have been unable to explain the operation mechanism of modern warfare, which is an urgent need to from the perspective of cognitive science explore the running mechanism of command mechanism, and the success of a better understanding of modern warfare. The research path and method selected in this paper is an attempt to analyze and study the scientific cognition of command mechanism, hoping to provide reference for the research in related fields.

References

1. Hu, X.F., Yang, J.Y., et al.: Research on Capability Analysis Method of System Combat System Based on Complex Network. Science Press, Beijing (2019)
2. Hu, X.F., He, X.Y., Rao, D.H.: Research on the analysis method of system combat coordination capability based on complex network. Complex Syst. Complex. Sci. **2**, 86–92 (2015)
3. Network Centric Warfare: Developing and Leveraging Information Superiority. CCRP, Washington, DC (1999)
4. Luo, X.S., Luo, A.M., Zhang, Y.H.: Military Information System Architecture Technology. National Defense Industry Press, Beijing (2010)
5. Khasawneh, M.T., Shearer, N., Rabadi, G., et al.: The information age combat model: a vision for A discrete event simulation approach. Int. J. Simul. Process Model. **12**(5), 429–435 (2017)
6. Sheffi, Y., Daganzo, C.F.: Computation of equilibrium over transportation network: the case of disaggregate demand models. Transp. Sci. **14**, 155–173 (1980)
7. Milo, R.: Network motifs: simple building blocks of complex. Networks **298**(5594), 824–827 (2002)
8. Lan, Y.S., MAO, S.J., Wang, H.: Structure Theory and Optimization Method of Command Information System. National Defense Industry Press (2015)
9. Cares, J.R.: An information age combat model. In: Proceedings of 9 International Command and Control Research an Technology Symposium (2004)
10. Moravoik, M., Schmid, M., Burch, N., et al.: DeepStack: expert-level Artificial Intelligence in no-limit Poker. Science **356**(6337), 508 (2017)

Confirmatory Factor Analysis of Place Attachment in Agritourism

Zhimin Li

Abstract According to relevant research and the characteristics of agritourism, this paper discusses that place attachment is more appropriate than the variable of tourist loyalty. In various constitutive models of place attachment, place identity and place dependence are selected as two main dimensions, and the questionnaire of place attachment is compiled by referring to the measurement items of similar questionnaires and combining the agritourism situation. Through confirmatory factor analysis, the components of place attachment in agritourism were verified.

1 Introduction

In tourism research, the related research on tourist loyalty borrowed from marketing once occupied a dominant position. Tourism products are also commodities in nature, but they have their unique attributes. Tourists only have the right to use tourism products, but have no ownership, so it is relatively more free to convert tourism products. In terms of tourists' travel motivation, seeking novelty is universal psychology, which does not tend to form long-term loyalty to a fixed tourism destination. In the field of tourism, to describe tourists' preference and a certain amount of loyalty to a destination, it may be more appropriate to use the concept of place attachment. By comparing various tourism destinations, tourists can identify or constantly strengthen the destinations which are relatively more recognized and liked, and then form place attachment. Place attachment can better express tourists' sense of belonging to destinations and a certain degree of elastic relationship between them. Destination loyalty emphasizes the implementation of actual tourism behavior and results, while place attachment can just be a tendency to repetitive tourism; Tourist loyalty requires the exclusiveness of the choice of the same kind of tourist destination, while place attachment mainly shows that a specific destination always occupies a prominent position in the recreation opportunity spectrum (ROS) of tourists, showing a higher frequency of patronage. The destination of place attachment is not unique, and each tourist

Z. Li (✉)
Business School, Sichuan University, Sichuan, China

L. C. Jain et al. (eds.), *3D Imaging Technologies—Multidimensional Signal Processing and Deep Learning*, Smart Innovation, Systems and Technologies 236,
https://doi.org/10.1007/978-981-16-3180-1_17

may have his destination selection sequence. Their significance to tourists' lies in constructing the "base" or "base camp" for tourists' leisure travel, and at the same time, tourists can explore new destinations outward. For agritourism, a tourism type with leisure as its main purpose, tourists still hope to catch novelty, because people all have an inherent instinctive demand of "new information stimulation," and tend to try new destinations to obtain more new stimuli and greater marginal utility. Compared with other types of tourism destinations, agritourism destinations are more likely to promote tourists to form place attachment. Therefore, it is necessary to explore and verify place attachment and its constituent elements in the context of agritourism.

2   Measurement Index of Place Attachment

Existing researchers generally agree that place attachment includes two aspects: Place dependence and place identity. Place dependence refers to the attachment to place functions, which can satisfy and realize people's specific needs and goals, and represents the level of utilitarian rational cognition. Place identity represents the level of emotion and symbolic meaning. The feeling of deep attachment is caused by positive emotions. The coordination between place image and tourists' self-cognition is an important source of symbolic meaning. Place dependence is substitutable, to form stable place attachment, place identity with symbolic meaning and deep emotional connection is needed, and place identity is closely related to affirmation of place function. Williams, McIntyre believed that tourists have an emotional connection with places not only because destinations have their preferred attributes, but also because they provide meaning, identity, and life goals. Some familiar landscape attributes would strengthen tourists' emotional connection, but this emotional connection is not an inevitable result of regular visits or long-term interaction [1].

Place attachment is finally reflected as a kind of holistic tendentiousness or nostalgia emotion, so some researchers recently think that place attachment should not be stratified, but considered as a unified latent variable or a single observed variable. Jorgensen, Stedman just integrated conative, affective, and cognitive into a single-place attachment scale. Therefore, the measurement items of place dependence and place identity can be directly incorporated into the unified latent variable of place attachment, and these two dimensions can be extracted when analyzing [2–4].

According to the above analysis, combined with the characteristics of agritourism, the measurement items of place attachment are selected. So far, the measurement contents of place attachment were listed in relevant studies. Although the measurement objects and environments are different, and the composition and structure are different, the specific measurement items are universal. This study draws lessons from these measurement items when constructing the measurement indicators of place attachment of agritourism [5, 6].

3 Questionnaire Design

Based on theoretical considerations, place attachment is set as a unified latent variable, which forms a first-order model with its constituent elements. There are also statistical reasons for this. Qiu Haozheng and Lin Bifang pointed out that based on the principle of model recognition, the number of first-order factors forming high-order factors should not be less than three, and otherwise it will cause insufficient recognition. Therefore, a simple algorithm defined by the HCFA model is that each high-order factor must be defined by three or more initial-order factors [7]. So this study did not choose to build a second-order model of place attachment, but took the measurement items of place dependence and place identity as the observation variables of place attachment directly. According to the relevant theories and referring to the measurement items of existing research, this paper designed six measurement questions about place attachment in the questionnaire, namely: i. I have benefited a lot here(Z1); ii. This place makes me linger more than other places(Z2); iii. It can meet my specific needs(Z3); iv. I have a sense of belonging here(Z4); v. I enjoy visiting here very much(Z5); vi. This place is of special significance to me(Z6). The questions are set options according to the 5-point Likert scale method, and 1 ~ 5 is used to correspond to the answer of each question, which indicates the degree of recognition of tourists to each measurement index. The transition from 1 to express strong disagreement to 5 to express strong agreement. To improve the representativeness of research data samples, this study selected four agritourism scenic spots in Chongqing for field investigation and data collection. A total of 248 questionnaires was distributed and 212 valid questionnaires were collected, with an effective rate of 85%.

4 Data Analysis

According to the questionnaire survey data of place attachment in agritourism, the structural equation model (SEM) was used for confirmatory factor analysis (CFA). The AMOS 21.0 software was used for confirmatory factor analysis, and the maximum likelihood estimation method (ML) was adopted. The initial fitting result is not ideal. According to the model modification index (MI), the covariance of e4 ← → e5 is added, and finally the place attachment structure model of agritourism is obtained, as shown in Fig. 1. The model can be estimated by convergence without negative error variance, which means that the parameters in the model have no unreasonable solution values. The specific values of the standardized regression coefficient (factor load) are all between 0.70 and 0.95 (see Table 1), indicating that the basic fitness of the model is good. According to the standardized regression coefficient and measurement error, the composite reliability (CR) and average variance extraction (AVE) of the latent variable (place attachment) can be calculated to test the intrinsic quality of the model. In the place attachment model, the composite reliability

Fig. 1 Confirmatory factor
analysis of place attachment

Chi Square=16.097
df=8
p=.041

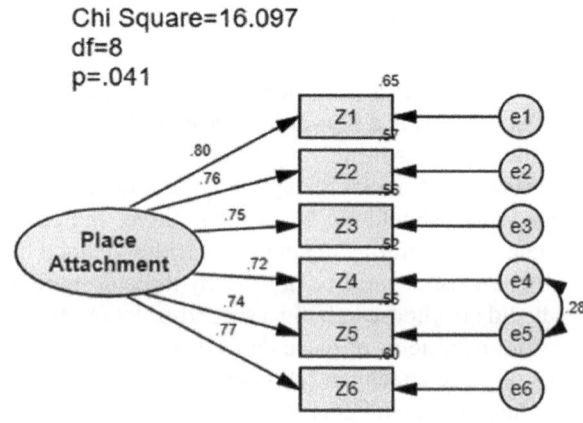

Table 1 Standardized
regression coefficient,
composite reliability, and
average variance extraction
value of place attachment

Measurement index	Factor load	CR	AVE
Z1 ← place attachment	0.804	0.889	0.572
Z2 ← place attachment	0.756		
Z3 ← place attachment	0.745		
Z4 ← place attachment	0.719		
Z5 ← place attachment	0.738		
Z6 ← place attachment	0.774		

Table 2 Fitting index of
confirmatory factor analysis
of place attachment in
agritourism (N = 212)

Leading indicator	Standard value	Fitted value
$\chi 2$		16.097
df		8
$\chi 2/df$	<3	2.012
NFI	>0.90	0.976
RFI	>0.90	0.954
NNFI	>0.90	0.976
CFI	>0.90	0.987
IFI	>0.90	0.988
GFI	>0.90	0.976
RMSEA	<0.08	0.069
SRMR	<0.08	0.025

is 0.889>0.6, and the average variance extraction is 0.572>0.5, so both composite
reliability and average variance extraction meet the requirements, indicating good
reliability and validity [8].

The fitting index of confirmatory factor analysis is shown in Table 2. The results show that the chi-square value ($\chi 2$) of the model is 16.097, the degree of freedom (df) is 8, and the ratio of the two is 2.012, which is lower than the standard value of 3.0. The fitness indexes of the model, including comparative fitting index (CFI), non-normal fitting index (NNFI), goodness of fit index (GFI), incremental fitting index (IFI), normal fitting index (NFI), and relative fitting index (RFI), all reach critical value of 0.9; In the approximate error-index, the root mean square error of approximation (RMSEA) and the standardized root mean square residual (SRMR) are both less than the standard value of 0.08. Taken together, the above indexes reflect that the model fits well with the data, which shows that the place attachment model of agritourism has an ideal fitting degree, thereby showing that the model has good structural validity.

5   Research Results and Conclusions

The results show that place attachment is suitable for measuring the close relationship between tourists and destinations in the context of agritourism. Once a place attachment is formed to a agritourism destination, although it may not be chosen to visit every time, the attached place will become a periodic main travel destination and play an increasingly important role in people's lives. From the measurement items, the factor load (0.804) of "I benefit a lot here" is the highest, which shows that it is very important for agritourism to bring comprehensive benefits to tourists. Of course, if a agritourism destination wants to be the object of attachment, it must have its characteristics, and "can meet my specific needs," which cannot be replaced by other places, and "makes me linger more than other places," thus forming a functional dependence. Based on place dependence, it brings tourists a sense of safety and comfort and then produces a "sense of belonging," which is a natural result. To gain recognition, agritourism scenic spots should also bring tourists positive emotions such as "enjoyment during vacation." The image of agritourism destination itself has symbolic significance, and if it has witnessed important or memorable events in tourists' lives, it has "special significance" for tourists. To sum up, place attachment is a psychological state of strong adhesion achieved by mingling tourists' positive cognition and emotion of destination, which is an important goal of agritourism destination development. It is worth noting that the perceptions of leisure agricultural scenic spots as "places" will be influenced by their located "region." Only by constructing good large and small environments can place attachment formed, which is a vivid reflection of the concept of tourism system development.

References

1. Brocato, E. D.: Place attachment: an investigation of environments and outcomes in a service context. Bus. Admin. (2006)
2. Jorgensen, B.S., Stedman, R.C.: Sense of place as an attitude: lakeshore owners attitudes toward their properties. J. Environ. Psychol. **21**(3), 233–248 (2001)
3. Prayag, G., Ryan, C.: Antecedents of tourists' loyalty to mauritius the role and influence of destination image, place attachment, personal involvement, and satisfaction. J. Travel Res. **51**(3), 342–356 (2012)
4. Ramkissoon, H., Weiler, B., Smith, L.D.G.: Place attachment and pro-environmental behavior in national parks: the development of a conceptual framework. J. Sustain. Tour. **20**(2), 257–276 (2012)
5. Ramkissoon, H., Smith, L.D.G., Weiler, B.: Testing the dimensionality of place attachment and its relationships with place satisfaction and pro-environmental behaviors: a structural equation modeling approach. Tour. Manage. **36**(36), 552–566 (2013)
6. Suntikul, W., Jachna, T.: The co-creation/place attachment nexus. Tour. Manage. **52**, 276–286 (2016)
7. Williams, D. R., Mcintyre, N.: Place affinities, lifestyle mobilities, and quality-of-Life. Handbook of Tourism and Quality-of-Life Research Springer Netherlands, pp. 209–231, Springer (2012)
8. Qiu, H.Z., Lin, B.F.: Principle and Application of Structural Equation Model. China Light Industry Press, Beijing (2009)

Application of Computer Sensor and Detection Technology in Mechatronics

Guojun Wang

Abstract In recent years, along with the rapid economic development of our country, in the motor have been living in China's industrial production, occupy the important position, is the indispensable equipment, the motor in case of problems, failure, not only can directly lead to the normal operation of mechanical system, but also will lead to serious economic loss for the companies, have more very person will endanger the life safety of the operator. Therefore, it is necessary to constantly improve the manufacturing process of the motor, expand the application range of the motor, and analyze and study the hidden dangers of the motor. In this article, the fault diagnosis of electrical equipment is mainly carried out by using computer sensors and related detection techniques. The possible failure factors of the motor are analyzed and studied, so as to detect the working state of the motor.

1 Research Status of Motor Fault Diagnosis

In recent years, as enterprises gradually tend to mechanized processing, China has gradually attached importance to mechanical maintenance and maintenance work. At the beginning, the enterprise adopted the method of monitoring the mechanical facilities and equipment, but the mechanical equipment still appeared faults, damages and other phenomena. For this reason, researchers have found through continuous research, combined with different technical theories in different fields, different testing methods are adopted for different mechanical equipment. Although there are many ways of detecting motor fault diagnosis, there are great differences between different ways of diagnosis.[1, 2] According to the continuous research and observation of researchers, the motor fault diagnosis process can be divided into the following four steps: Data acquisition, data preprocessing, feature extraction and fault classification (Fig. 1). The first step of motor fault analysis is data acquisition. The data acquisition will directly determine the motor fault type. Generally, current signals or vibration signals are used for diagnosis. When the motor fails, the amplitude

G. Wang (✉)
Water Conservancy of Shandong Technician College, Shandong, China

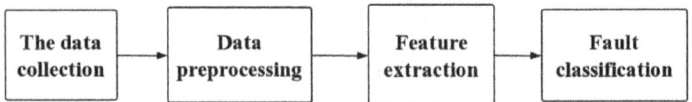

Fig. 1 Motor fault diagnosis steps

and frequency involved in the motor will change. This is to judge the motor failure through different vibration frequencies. The other way is to judge the fault type of the motor by the change of current when the stator current in the motor changes significantly. Data preprocessing and feature extraction are mainly used to analyze faults by eliminating noise. Half of them adopt non-stationary signal processing mode, which has been widely used in fault signal feature extraction and fault detection. In recent years, with the continuous development and progress of mechanical detection technology, the requirements on the technological level are more and more strict, so as to continuously improve the acquisition of signal accuracy and other issues. When the vibration received signal becomes stronger and stronger, many important information will be received. If the noise interfered signal becomes stronger and stronger, the amplitude of vibration received signal will be weak, and the detector will be unable to recognize the signal.[3–5]

2 Motor Fault Type

In this paper, common fault types of motor are summarized, and three types of motor fault types in Table 1 are analyzed.

2.1 Bearing Failure

Bearing fault is a common type of motor fault, the fault is divided into two types: Outer ring fault and inner ring fault.[6–8] In this paper, the outer ring fault is analyzed and studied. See Table 2. When the rolling bearing fails during operation, a complex

Table 1 Motor fault types

The fault types	The fault details
Normal	\
Motor bearing fault	Bearing outer ring fault, see groove scar marks
Motor turn-to-turn short circuit fault	There is a short circuit between turns, and the copper wire between turns shows that the insulation skin is broken
Motor rotor bar break fault	Rotor broken bars, see the signs of the rotor broken bars

Table 2 f_{OD} parameters

f_m	The rotation frequency of the motor
n	Number of rolling objects
B_D	Rolling diameter
P_D	Bearing pitch diameter
β	The angle of contact of the rolling body

vibration signal will be generated. The main reason for the failure is that the bearing parts are installed or processed by external forces, so that the rotor produces asymmetric force, resulting in the rotor shaft wear, when it reaches a certain degree, the bearing has a fault.

The diagnosis method of motor bearing can usually be calculated according to the vibration frequency of the outer ring. The specific vibration frequency formula is as follows:

$$f_{OD} = \frac{n}{2} f_\infty \left(1 - \frac{B_D}{P_D} \cos \beta \right)$$

2.2 Inter-Turn Short Circuit Fault

Inter-turn short circuit fault is also a common fault of the motor. Inter-turn short circuit will be easily produce super large current, which will cause the motor to burn out. When the polar logarithm of the motor is P, if the current is too large, it will directly cause a circuit short circuit, which will cause the coil inside the motor to burn out, change the internal magnetic field structure of the motor, and reduce the motor torque. [9, 10]Therefore, inter-turn short circuit has a great influence on motor fault and is the key fault type to prevent. There are many reasons for the short circuit fault of the motor, which can be generally divided into the following four types:(1) During the switching process of the motor, the inter-turn insulator bears the voltage instantaneously. (2) The electrical temperature of the motor is too high, and the electrical conductivity of the insulator is enhanced. (3) The outer winding of the motor stator produces vibration, thus causing damage to the insulator. (4) The motor is damp or operates under high-temperature environment. In the case of inter-turn short circuit fault of the motor, a current loop will generally form, and the current loop will form a magnetic field between the stator and rotor of the motor. The magnetic field formula is as follows:

$$f = F_m \cos(\omega t) \cos(p\theta)$$

By analyzing and studying the vibration direction and amplitude between the stator and rotor in the electron, it can be found that the vibration direction and

Table 3 F_1 parameter

K_1	A constant associated with the polar logarithm
N_1	Number of windings per phase of stator
ω	The fundamental angular frequency of the current
P	The polar logarithm of the motor
θ	Stator mechanical angle
I_1	The stator current

amplitude change when affected by the motor fault. Therefore, vibration data can be used to detect the fault of inter-turn short circuit.

2.3 Rotor Bar Breaking Fault

When the motor rotor bar fault occurs, mainly refers to the rotor due to some unexpected damage, so that the motor cannot operate normally. The main reasons for rotor slip-breaking faults are rotor manufacturing process problems or motor switching on and off several times in a short period of time. See Table 3 for details. When the motor is in the process of operation, the rotor breaks under the action of some external force. In order to make the rotor run normally, the electronics increase the power. In this way, the stator cannot drive the rotor and the motor cannot be used normally.

When the motor appears the fault of broken rotor, it is usually detected through the current, the fixed rotor of the motor can be combined with each other through a certain magnetic field, when the motor appears the fault of broken rotor, the stator current will change, so the general detection personnel through the current to detect the fault of broken bar. The specific stator detection formula is as follows:

$$F_1 = K_1 N_1 I_1 (\omega t - p\theta)$$

When the motor fails, the motor will not start. Therefore, the motor can be diagnosed and detected through current or vibration signal and other factors.

3 Conclusion

In this paper, the common fault types of the motor are analyzed and studied, and the detection method is proposed when the motor fault occurs. To sum up, the fault types involved in this paper are all the study data through vibration signal or current.

References

1. Mohammed, Z., Ángel, R.: Micro-electromechanical sensors in the analytical field. Analyst. **134**(7), 1274–1290 (2009)
2. Hur, S., Lee, K.H., Han, Y.B.: High sensitivity electromechanical sensor based on ZnO Nanowire FET on flexible substrate. Theor. Comput. ence. **350**(1), 40–48 (2006)
3. Park, J. H., Kwon, T. Y., Yoon, D. S., Kim, H., Kim, T. S.: Fabrication of microcantilever sensors actuated by piezoelectric $Pb(Zr_{0.52}Ti_{0.48})O_3$ thick films and determination of their electromechanical characteristics. Adv. Funct. Mater. **15**(12), 2021–2028 (2010)
4. Han, Y., Wang, G., Yu, J., Liu, C., Zhang, Z., Zhu, M.: A Service-based approach to traffic sensor data integration and analysis to support community-wide green commute in China. IEEE Trans. Intell. Transp. Syst. **17**(9), 2648–2657 (2015)
5. Durrant-Whyte, H.F.: Sensor models and multisensor integration. Int. J. Robot. Res. **7**(6), 97–113 (1988)
6. Luo, R.C., Kay, M.G.: Multisensor integration and fusion in intelligent system. IEEE Trans. Syst. Man Cybern. **19**(5), 901–931 (1989)
7. Lin, Y.D., Ko, M.C., Wu, S.T., Wu, W.T., Hsieh, P.C.: Abstract 16797: anisotropic cardiac patch promotes electromechanical integration of transplanted cells and therapeutic efficacy after infarction. Circulation **2012**(21), A16797 (2012)
8. Kehat, I., Khimovich, L., Caspi, O., Gepstein, A., Shofti, R., Arbel, G., Huber, I., Satin, J., Itskovitz-Eldor, J., Gepstein, L.: Electromechanical integration of cardiomyocytes derived from embryonic stem cells. Nat. Biotechnol. **22**(10), 1282–1289 (2004)
9. Morris, J.C.: A simple electromechanical sensor for positioning and holding applications (Notes). Rev. Entific Instrum. **41**(6), 877–878 (1970)
10. Reiche, C.F., Körner, J., Büchner, B., Mühl, T.: Introduction of a co-resonant detection concept for mechanical oscillation-based sensors. Nanotechnology **26**(33), 335501 (2015)

Establishment of Smart Customer Service System Based on Data Mining

Qingqing Ren, Wanzhen Ma, Liu Yang, and Lifei Yan

Abstract With the continuous upgrading of mobile Internet technology and the continuous innovation of business models, telecom operators gradually carry out Internet transformation. With the in-depth reform of Internet-based operators, new challenges are presented to the window customer service directly facing the end users. The service awareness related to the diversity of users' demands, the differentiation of demands information, and the timeliness and accuracy of service requirements has been significantly improved. However, it is necessary to pay attention to customer service management, keep up with the pace of customer demands, improve internal operation management, and meet the improvement of service quality. In this paper, from the perspective of the problems and challenges faced by the development of customer service business, based on data mining technology, using speech recognition engine and text processing technology, combined with the pain points and demands faced by the current customer service business, the intelligent customer service operation management business application based on data mining is proposed. This will support the internal operation management to improve quality and efficiency, focus on customer service, and provide hot information discovery. The intelligent customer service system is constructed through data mining comprehensive model algorithm thematic analysis, so as to satisfy customers' aspirations, focus on hot issues and demands, better serve customers, and improve customer service quality.

1 Introduction

With the development of Internet products and related businesses, Internet products put forward higher requirements in customer service system. All kinds of

Q. Ren (✉) · W. Ma · L. Yang
State Grid Xinjiang Electric Power Co., LTD, Xinjiang, China

L. Yan
Fujian Yirong Information Technology Co., LTD, Fujian, China

© The Author(s), under exclusive license to Springer Nature Singapore Pte Ltd. 2021   151
L. C. Jain et al. (eds.), *3D Imaging Technologies—Multidimensional Signal Processing and Deep Learning*, Smart Innovation, Systems and Technologies 236,
https://doi.org/10.1007/978-981-16-3180-1_19

Internet-based products have shown the characteristics of form innovation, strong personalization, small products, and quick iteration, aiming to respond and change quickly around the needs of users. Therefore, the service team should have the ability of professional service and rapid deployment of business. Under the condition of Internet products, users call more frequently, and for longer periods, the number of complaints fluctuates greatly, and the derogatory mention rate is higher. Therefore, the intelligent customer service system should build a customer service collaboration mechanism with internal horizontal collaboration, vertical penetration, and internal and external collaboration [1, 2]. Data mining belongs to a new research field of machine learning. It is a feature learning method, which mainly explains data by establishing human brain simulation mechanism. It can also be used for unstructured data. Deep learning has a wide range of applications, which can be applied to images, sounds, and texts (Note: the intelligent customer service application mentioned in this paper belongs to sound applications) [3–6]. Data mining methods can be divided into supervised learning and unsupervised learning. Data mining transforms the original data into a higher-level and more abstract expression learning with a simple, nonlinear model. The foundation of data mining is that it can learn very complex functions through enough transformations and combinations and make them play an important role in contemporary society. Because the traditional service system cannot meet the service requirements of Internet products, a solution is urgently needed to solve this problem. By using the popular technologies and combining the service innovation concept of the communication industry, we can satisfy the customer service and operation based on Internet, realize the support of professional, big data, and intelligent management system, escort the customer service, and help the strategic transformation of the new IT ecology. Based on data mining and data flow monitoring, this paper constructs the general idea and implementation scheme of intelligent customer service system for telecom operators [6, 7].

2   Methodology

Neural network data mining is a kind of machine learning architecture, that is, algorithm, in which all individual units are connected together in the way of weights, and these weights are obtained by training network. Data mining has also made good achievements in natural language understanding, especially in the following aspects: topic classification, automatic question answering and language translation, and emotion analysis. It can find potentially complex structures in big data by using BP algorithm. BP algorithm, i.e., backward propagation algorithm, can guide the machine how to use the previous layer to obtain errors and adjust the internal parameters of the current layer, which is outstanding in speech, audio, and video, while recursive network is slightly superior in processing sequence data, such as text analysis and speech recognition [8–10].

The layered output of data mining can be defined as O_j:

$$O_j = G_i(F) = \exp\left(-\frac{\|F - t_j\|^2}{\sigma_j^2}\right)$$

F is the input vector, t_j is the central value vector of the hidden layer, σ_j^2 is its width, and j is the neuron. The actual network output is:

$$y_k = \sum_{j=1}^{n} o_j w_{ij}$$

where n is the number of hidden layer nodes.

In addition, the error function e can be defined as (k is a constant):

$$E = \frac{1}{2} \sum_{k=1}^{k} (d_k - y_k)^2$$

Combining the least sparse mean clustering algorithm with the traditional least square method can improve the computational efficiency.

When the input signal is;

$$x_k(i = 1, 2, \cdots k)$$

The output of the hidden layer is:

$$h(j) = h_j\left(\frac{\sum_{i=1}^{k} w_{ij} x_i - b_j}{a_j}\right) \quad j = 1, 2, \cdots, L$$

h_j represents the JTH node output of the hidden layer, w_{ij} represents the connection weight of the input layer and the hidden layer, b_j represents the translation factor of the wavelet basis function, a_j represents the expansion factor of the wavelet basis function, and the calculation formula of the output layer of the wavelet neural network is as follows:

$$y = \sum_{i=1}^{k} w_{jk} h(i) \quad k = 1, 2, \cdots, m$$

Calculate other input samples according to the above formula to obtain the structure diagram 2 of the perception machine, where x is the customer service voice sample, a is the path, and h is the last input place.

3 Evaluation and Analysis

In this paper, the data of operators' order data containing a large number of complaint texts are collected, and how to quickly collect the main contents, that is, quickly classify these complaint texts, and get timely solutions to various complaint problems. This requires a reasonable classification of complaint texts. In this experiment, a total of 12,600 pleadings were selected. The length of each pleadings body was 1–100, and the contents of the pleadings were different (Fig. 1).

Table 1 shows the results of the test set obtained by each machine classification through data mining algorithm and training model. The accuracy rate, recall rate, and F1 value are used in the table to measure the classification effect. It can be seen from the table that the accuracy rate and recall rate of all algorithms are about 0.5, which shows that the intelligent customer service system established by the data mining algorithm selected in this paper has positive effects.

Table 2 is the overall effect evaluation. It can reflect the simultaneous hit rate of the classification algorithm we use. The overall evaluation function gives two evaluation indexes: coverage and recall. Coverage represents the percentage of documents that have reached the recall threshold. A coverage is defined as follows: Coverage = nowhere k represents the number of algorithms that meet the threshold, and n

Fig. 1 Fluctuation diagram of customer complaints after applying smart customer service

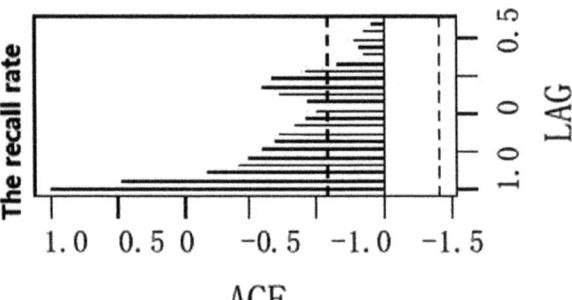

Table 1 Algorithm precision results

Accurate rate	Recall rate	F1 value
0.524	0.434	0.434
0.431	0.442	0.431
0.551	0.524	0.512
0.461	0.452	0.436
0.469	0.431	0.419
0.451	0.451	0.432
0.418	0.399	0.399
0.382	0.361	0.307
0.479	0.470	0.459

Table 2 Overall effect evaluation results

Overall effect evaluation	n_ensemble coverage	n_ensemble recall
n > = 1	1	0.46
n > = 2	1	0.46
n > = 3	1	0.46
n > = 4	0.94	0.47
n > = 5	0.76	0.54
n > = 6	0.67	0.6
n > = 7	0.52	0.67
n > = 8	0.38	0.76
n > = 9	0.15	0.93

represents the total number of algorithms. It can be seen from Table 2 that when the total number of algorithms is small, the coverage is large and the recall rate is low. When the total number of algorithms reaches 9, the effect is already very good, and a low percentage of documents can achieve a high recall rate, it can be seen that the fluctuation trend is gradually decreasing, which shows that the intelligent customer service system has a positive role in promoting business, and that the algorithm selected based on data mining is effective.

4 Conclusion

With the continuous development of intelligent technology, the common mission of major telecom operators has become how to introduce new technologies to serve customers and keep continuous innovation. Therefore, operators should continue to research and build intelligent customer service systems, so that users can enjoy convenient services through data mining, machine learning, and other information technologies, so that intelligent customer service systems can have the ability of man–machine interaction without asking the prophet.

References

1. Soelberg, E. L., Berry, M. J.: System and method for using cellular network components to derive traffic information. US Patent 8,494,496,(2013)
2. Ghosh, D.: Apparatus, method, and computer program product for extracting insights from bill presentment data. US Patent App. 14/496,572, (2016)
3. Triola, R. C.: Method, system, apparatus, and program for displaying targeted advertisements. US Patent App. 13/452,706, (2012)

4. Gebremeskel, G.B., Yi, C., Wang, C.L., He, Z.S.: Critical analysis of smart environment sensor data behavior pattern based on sequential data mining techniques. Ind. Manag. Data Syst. **115**(6), 1151–1178 (2015)
5. Chuan, W., Suping, P.: The establishment of a trap information system based on GIS. Comput. Geosci. **29**(7), 937–940 (2003)
6. Zhang, L., Zhao, H., Li, Q.S., Wang, J., Tan, X.: Establishment of paper assessment system based on academic disciplinary benchmarks. Scientometrics **84**(2), 421–429 (2010)
7. Wang, C., Peng, S.P.: The establishment of a trap information system based on GIS. Comput. Geosci. **29**(7), 937–940 (2003)
8. Min, B.W., Cho, H.J.: Improvement of information service system for smart library based on bigdata. Development **128**(20), 4057–4067 (2013)
9. Karger, T., Karger, S.: Some thoughts from your publisher. Pharmacology **65**(1), 1–1 (2002)
10. Schallehn, H., Seuring, S., Strähle, J., Freise, M.: Customer experience creation for after-use products: a product-service systems-based review. J. Cleaner Prod. **210**(FEB.10), 929–944 (2019)

Application and Development of Natural Language Processing Service in Intelligent Customer Service System

Yajie Li, Tao Wang, Shuting Chen, and Xiaodong Zhang

Abstract With the arrival of the era of big data, various new concepts such as network technology and artificial intelligence are gradually integrated into people's daily life, and their convenience and efficiency can be seen. As far as enterprises are concerned, the traditional customer service no longer meets the needs of the development of the times. The establishment of intelligent customer service system reduces the working pressure of manual customer service while improving the working efficiency. Save the cost of the enterprise, so the related technology is constantly improving. In the intelligent customer service system, natural language processing service is a core technology, and it is the key to realize the intelligent customer service system to communicate with customers like artificial customer service. This paper introduces the key technologies of intelligent customer service system, namely speech recognition, machine learning, and natural language processing, and then introduces the application status of natural language processing service technology in intelligent customer service system.

1 Introduction

Of more common customer service system can be roughly divided into three stages: pre-sale, sale, and after-sale. Enterprises invest a lot of manpower, material and financial resources in the process of customer service system construction. Ring by ring, one-ring problems may result in the follow-up work to the wrong direction, and error correction process is also laborious and increase the cost. If artificial intelligence natural language processing technology is introduced, this situation will be greatly improved, which can not only improve the service efficiency of customer service but also improve the quality of customer service.[1–5] Natural language processing

Y. Li (✉) · T. Wang · S. Chen
State Grid Xinjiang Electric Power Co., LTD, Wulumuqi, Xinjiang, China
e-mail: jie.15@163.com

X. Zhang
Fujian Yirong Information Technology Co., LTD, Fuzhou, Fujian, China

© The Author(s), under exclusive license to Springer Nature Singapore Pte Ltd. 2021
L. C. Jain et al. (eds.), *3D Imaging Technologies—Multidimensional Signal Processing and Deep Learning*, Smart Innovation, Systems and Technologies 236,
https://doi.org/10.1007/978-981-16-3180-1_20

technology is a technology, in which computers can learn, understand and generate human language according to the results of speech recognition, and perform intelligent processing according to the results of language understanding. To put it simply, natural language processing is mainly composed of two parts. One is natural language understanding; that is, understanding each word and gradually analyzing the nature of the word, whether it is a verb or a noun. The second is natural language generation; that is, through the use of machine learning, further deep learning to complete the transformation of computer language into a language that ordinary people can understand. The purpose of this paper is to analyze the application and development of natural language processing service technology in the intelligent customer service system, so as to scientifically add artificial intelligence processing content into the original customer service system. As shown in Fig. 1, the work efficiency and quality of the entire customer service system can be improved by improving the structure and content of the entire customer service system.

In the new system architecture, speech recognition and speech synthesis modules replace the original mode of human communication with users. Natural language is used to process the corresponding phone call or work order instead of the manual operation previously undertaken by the business processing personnel. The knowledge management module is used to replace the original knowledge base construction method. Artificial intelligence technology module replaced the original manual processing, with a certain repetitive and regular work. The application of the new system can reduce labor costs and improve efficiency and effectiveness without affecting customer service quality [6].

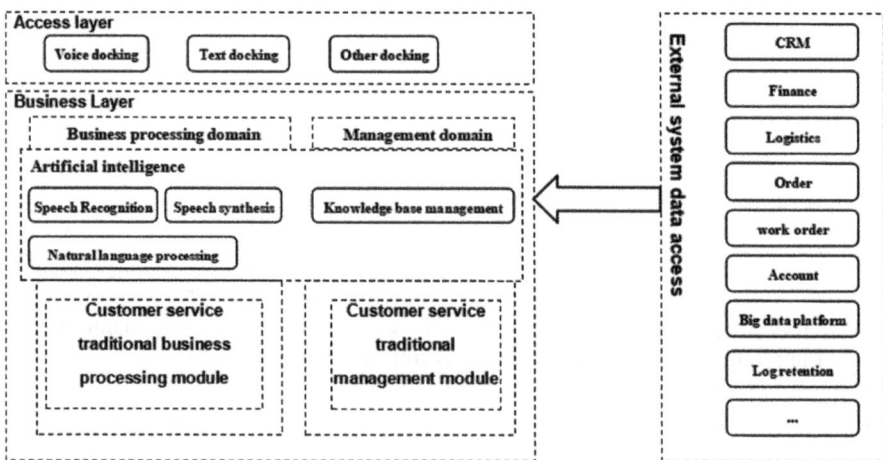

Fig. 1 Schematic illustration of customer service system architecture incorporating natural language processing technologies

2 Methodology

For practical applications, the matching technique of association recommendation for natural language processing services usually uses Pearson correlation coefficient and vector angle cosine similarity to calculate the matching degree. A method for calculating linear correlation, also known as product difference correlation coefficient or product moment correlation coefficient, is formally defined as:

$$\text{Sim}(u; v) = r \frac{s2 S_{uv}(r_{u;s} i \overline{r_u})(r_{v;s} i \overline{r_v})}{s2 S_{uv}(r_{u;s} i \overline{r_u})^2 s2 S_{uv}(r_{v;s} i \overline{r_v})^2}$$

where the SIM represents the language processing feature, Suv represents the acquired vector, r represents the cosine similarity. However, considering the influence of various attributes of users on the calculation of matching degree, it can be seen from the customer model that different customer attributes are not isolated, but there are various connections. Therefore, based on the above basic formulas, a matching degree calculation method is proposed in this section, which considers a variety of customer attributes and the relationship between customer attributes:

$$S(o; o^0) = \frac{B}{jN \ (o)jjN \ (o^0) \ j}$$

where S means connection weight, J means bias.
$B \in [0,1]$ Is a constant factor.
$N(O)$ A set of autocorrelation vertices representing the o of an object;
$|N(O)|$ Represents the potential of the set.
If the complex implementation process of natural language processing technology is considered, that is, when the user obtains an answer given by the online customer service, the customer has an evaluation feedback on the customer service. If the score is high, positive feedback is formed to increase the weight, whereas negative feedback is formed to reduce the weight [7]. In order to simplify the calculation without affecting the results, it is considered that the value of the matching reliability is related to the specific gravity of the four components in the customer model, that is:

$$W_i = \frac{S_i}{S_i + S_b + S_v + S_s}$$

The S_i attribute matching degree, the S_i behavior attribute matching degree, the S_v importance degree matching degree, the S_s social attribute matching degree, the confidence weight, the customer service system will make the intelligent speech recognition and the response according to the W_i confidence weight.

Partial rule representation of natural language processing techniques (Table 1):
Basic forms of generative systems are shown in Table 2

Table 1 Rule representation

(defrule < rule name > [< comment >]
< patterns > : Left—Hand Side (LHS) of the rule
= = >
< action > ;
Right—Hand Side (RHS) of the rule

Table 2 Basic generation systems

If (premise 1, premise 2,-, premise n) then (conclusion)
If (scenario, premise 2, premise n) then (conclusion)
If (scenario, premise 2, premise m)

To expand the premise of rules and limit the premise of rules from the scope of application is to use different knowledge representation methods in different scenarios and stages. There are many scenarios in online customer service intelligent reasoning system, including intelligent matching reasoning and intelligent search algorithm. The conclusion of the rule can also be extended by the corresponding calculation function in different scenarios [8].

3 Evaluation and Analysis

The intelligent reasoning system of online customer service has a high requirement for human–computer interaction, so the interface between the intelligent reasoning machine and online customer service needs to be established. Online customer service intelligent inference machine uses forward reasoning mechanism, build inference engine based on natural language processing, reasoning is the first scene set by reasoning on the basis of the context information, and then from the working memory query calls related to the initial fact, finally obtain the corresponding rules of inference, form a conclusion and interpretation of information. The forward reasoning machine algorithm flow of online customer service intelligent reasoning system is shown in Fig. 2.

Intelligent customer service system intelligent speech synthesis example schematic Fig. 3.

In the actual operation, for these problems with low similarity, there must be human customer service intervention, and these results also provide certain solutions and directions for customer service. Based on natural language processing technology, the system is designed and implemented, and the correlation between intelligent reasoning knowledge and reasoning mechanism of online customer service is realized, which improves the flexibility, generality, and expandability of the system. The system mainly considers the design of knowledge base and the design of reasoning. In the knowledge base design, based on the characteristics of the system and the natural language processing confidence association knowledge, an extended

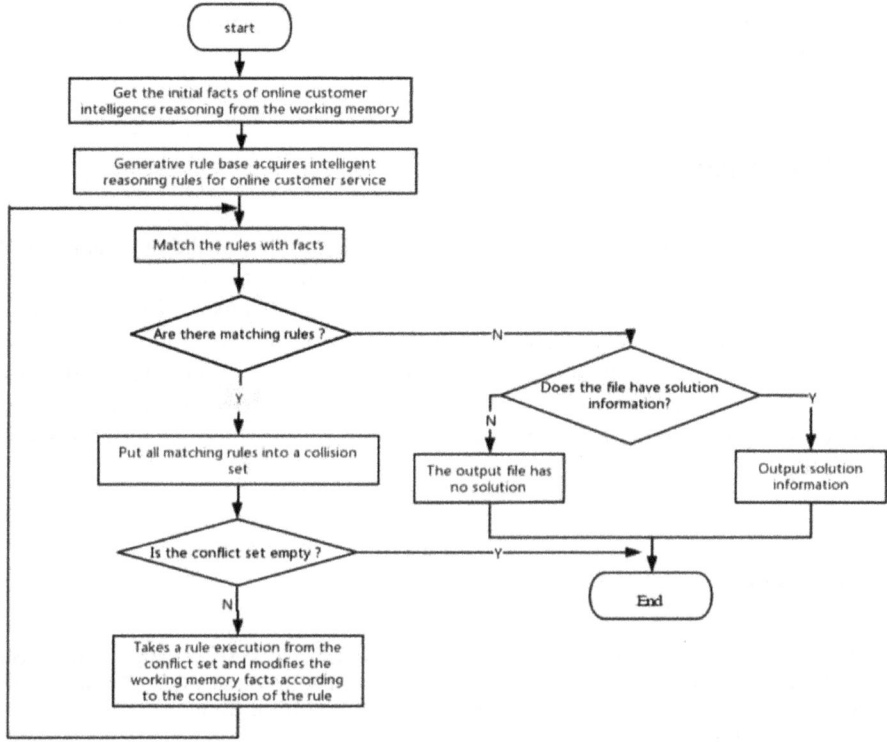

Fig. 2 Flow of forward reasoning machine for online customer service intelligent reasoning system

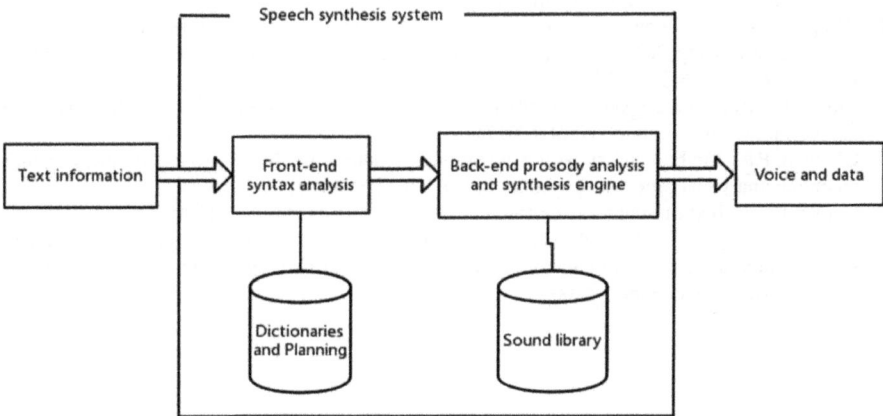

Fig. 3 Schematic diagram of intelligent speech synthesis system

Chinese production rule model is proposed. The reasoning machine of intelligent customer service system is designed.

4 Conclusion

Intelligent customer service system is the inevitable product of the development of big data era, which not only saves the cost of enterprises but also improves the efficiency of enterprises. Starting with the introduction of the key technologies of intelligent customer service system, this paper further analyzes the necessity of intelligent customer service system, which can help readers better understand the significance of intelligent customer service system. Intelligent customer service system will show a trend of popularization, and more functional and practical, broader development prospects.

References

1. Whittaker, S., Scahill, F., Attwater, D., Geenhow, H.: Practical Issues in the Application of Speech Technology to Network and Customer Service Applications, Interactive Voice Technology for Telecommunications Applications, Ivtta 98 IEEE Workshop. IEEE, Torino (2000)
2. Kusnadi, A., Antares, A. E.: Shallow Parsing Natural Language Processing Implementation for Intelligent Automatic Customer Service System, ICACSIS 2014. IEEE, (2014)
3. Rui, X., Ma, X.N., Li, P., Yang, L.B.: Nature language processing techniques and its applications in intelligent railway. Railway Comput. Appl. (2018)
4. Wang, Y., Sun, Y., Chen, Y.: Design and research of intelligent tutor system based on natural language processing. 2019 IEEE International Conference on Computer Science and Educational Informatization (CSEI). IEEE, (2019)
5. Cui, W.D.: Research on wechat intelligent customer service system based on natural language processing. Digital Technol. Appl. (2017)
6. Gladden, P.E.: Method and apparatus for providing neural intelligence to a mail query agent in an online analytical processing system: 08/646216, US (1998)
7. Gemer, t J.V.: Text mining tools on the Internet. ISIS technical report, (2000)
8. Jing, Y., Zhong, S., Li, X. et al.: Using instruction strategy for a web-based intelligent tutoring system. International Conference on Technologies for E-Learning and Digital Entertainment. Springer, Berlin, Heidelberg (2006)

Research on University Network Ecological Governance Based on Block Chain Technology

Peng Duan and Ying Li

Abstract As the main force of receiving and disseminating new knowledge and technology, the network platform has become the most important transmission medium now. There are hundreds of network platforms competing for glory, and various researches and opinions collide with each other and transmit the mainstream ideas. However, with the virtuality and freedom of the network, there are also many negative energy and bad information. The reasons can be summarized into three aspects: The information is not transparent, the network technology is not perfect, and poor supervision by the competent authorities. This paper will focus on this breakthrough point, take colleges and universities as the practice field, try to build "block chain + network ecological governance," innovatively put forward the implementation path of theory + technology, solve the outstanding problems of college network ecological governance at this stage, and provide technical reference for future related research.

1 Introduction

1.1 Concept of Network Ecology

Network ecology refers to the dynamic process in which users in cyberspace use their resources to transmit information through media in a specific structure and environment. In the ritual view of communication, communication can form a common collective representation and maintain a cultural system. Chen Lidan, Yi Zhenglin, and others (2009) analyzed the concept and category of network ecology in the book key words of communication, which provided a solid theoretical basis for related research [1].

P. Duan · Y. Li (✉)
Shandong Technology and Business University, Yantai, China
e-mail: 202013815@sdtbu.edu.cn

© The Author(s), under exclusive license to Springer Nature Singapore Pte Ltd. 2021   163
L. C. Jain et al. (eds.), *3D Imaging Technologies—Multidimensional Signal Processing and Deep Learning*, Smart Innovation, Systems and Technologies 236,
https://doi.org/10.1007/978-981-16-3180-1_21

1.2　Review of the Previous Research on University Network Ecological Governance

Through the academic retrieval and statistics of the research results in the past five years, more than 100 researches have been carried out around the theory and practice of university network ecology. Among them, Jing Juhua and Han Kaili (2015) conducted research on the construction of university network public opinion ecology, [2] Huang Junlu (2016) conducted empirical research on University Network Ecology and mainstream ideology communication,[3] Du Pengju (2018) conducted research on the construction of University Network Ideological and political education ecosystem under the environment of media convergence, and [4] Li Liangliang (2020) pointed out that the university should deal with the problems in university network public opinion ecology. In the process of campus network public opinion and maintaining campus security and stability, network opinion leaders play an important role. This paper studies the implementation path of creating "micro classroom" and "micro platform" of Ideological and political education [5]. Most scholars closely combine the construction of university network ecology with the requirements of university ideological work, and actively carry out theoretical research from multiple perspectives of university network culture.

1.3　Generation and Development of Block Chain Technology

The birth of block chain should be the most unusual and mysterious technology in the history of human science, because apart from block chain, so far, there is no major invention in the history of modern science, and no inventor can be found. On October 31, 2008, the founder of bitcoin, Tsung Nakamoto (pseudonym), published his paper "bitcoin: a peer-to-peer e-cash system" in the cryptography e-mail group. In this paper, the author claims to have invented a new set of electronic money system which is not controlled by the government or institutions, and the block chain technology is the basis of supporting the operation of bitcoin.[6] In January 2009, Zhongben Cong released the application case of block chain, the open-source software of bitcoin system, on the source. With the vigorous development of bitcoin, the research on block chain technology also began to show a blowout growth.

1.4　Current Application of Block Chain Technology

Many domestic scholars have also carried out relevant research according to China's national conditions. Huang Junfei and Liu Jie (2018) reviewed the block chain technology [7]. Dong Miao (2020) carry out practical research on blockchain and daily management of colleges and universities [8], special fund management and intelligent financial system architecture, involving many cores work of colleges and universities.

Through the above argument, it is not difficult to find that "Research on university network ecological governance based on block chain technology" has good academic value and application value. Among them, the academic value is reflected in: First, it helps to promote the high-quality development of the comprehensive management of colleges and universities, so as to consolidate the foundation of running colleges and universities; second, it innovates the ways and methods of network ecological governance of colleges and universities, fills in the gaps in the field, and enriches the application connotation of "block chain + colleges and universities"; the application value is reflected in: First, it is closely related to the requirements of the overall layout of the ideological work of colleges and universities implemented by the Party Central Committee. Each important link and node of university network ecological governance should focus on solving the pain points and difficulties. The second is the work model of block chain, which can realize the theory landing to a certain extent through optimization learning, and give reasonable suggestions for the network ecological governance of colleges and universities.

2 Implementation Path of University Network Ecological Governance Based on Block Chain Technology

Focus on the construction of "block chain + university network ecological governance," follow the idea of "network ecological ontology description—network ecological path optimization—block chain technology relearning—fusion analysis—model construction—model implementation—innovation and development." Taking the network ecological governance in colleges and universities as the research implementation object, focusing on the important deployment requirements of the ideological work in colleges and universities at this stage, this paper analyzes the prominent problems faced by the network ecological governance in colleges and universities, and expounds the essential reasons of the prominent problems. On this basis, by screening the advantages of block chain technology, it deduces the effective integration of block chain technology and network ecological governance in colleges and universities, so as to construct the network ecological governance to build an effective implementation model and provide innovative strategic support for university network ecological governance through practice. The research is carried out in five stages.

2.1 Research on the Current Situation of University Network Ecology in the New Era

In this stage, combined with the latest work requirements of national and local governments on the network ecological governance of colleges and universities, the relevant policies and regulations issued by governments at all levels, through combing the literature, consulting network information, questionnaire survey, interviews, and other research methods. We will select different levels of national representatives from the aspects of network ideology, network security, network environment construction, network information dissemination, and so on. In order to fully explore the current situation of university network ecology in the new era, especially the exposed prominent problems and challenges, factor analysis and multidimensional scale analysis are used to analyze university network ecology, representative events, and representative public opinion. The results of the research and analysis are formed into data, and the characteristic database is established to lay the sample research foundation for the follow-up research.

2.2 Research on the Optimization Path of University Network Ecological Governance

In this stage, focusing on the relevant requirements of the regulations on ecological governance of network information content issued by the State Internet Information Office, through inductive comparison method, the actual needs of network security maintenance, digital copyright protection, accelerated experience of network resource services, transparency of asset information and other aspects in colleges and universities are sorted out from the aspects of network ecological hardware and software, respectively; from the level of teachers and students, the actual needs of network security maintenance, digital copyright protection, accelerated experience of network resource services, transparency of asset information and other aspects are sorted out. This paper discusses the actual needs of teachers and students in many aspects, such as the integrity file management of teachers and students, the information privacy protection of teachers and students, the information transparency of awards and awards, the early warning and supervision of network public opinion, the construction of teachers' ethics, the Ideological and Political Education, the capital supervision, the daily behavior analysis, and the supervision of teachers and students' mental health. This topic will obtain the first-hand data through the questionnaire survey to evaluate the specific needs of the university network ecological governance. In the process of evaluation, the correlation coefficient and deviation degree are used to analyze the correlation of each link of the university network ecological governance, and finally sort out the optimization path of the university network ecological governance in the emerging stage.

2.3   Recognition of Block Chain Technology

In this stage, we will first analyze the knowledge map of the relevant literature, visit the block chain technology companies or bases, and use scientometrics to conduct word frequency analysis, CO word analysis, CO citation analysis and multivariate statistical analysis, so as to objectively sort out the technological frontier and development direction of block chain, and further establish "block chain +" as the subject research the guiding concept of core technology. Secondly, we should fully grasp the five characteristics of block chain: decentralization, openness, tamper-proof, anonymity, and traceability, and learn to understand the triple significance of block chain technology in today's society, that is, economic significance, social significance, and political significance. Finally, on the understanding of the disadvantages of block chain technology, this topic will clearly depict the risk factors of block chain technology at this stage from four aspects: immature technology development, new regulatory challenges, inapplicability of laws and regulations, misunderstanding of social cognition, and through MATLAB. The software uses the calculation formula of correlation degree to calculate the level of key early warning factors and provides relevant reference for the construction of the next "block chain + university network ecological governance" model.

2.4   Construction of University Network Ecological Governance Model Based on Block Chain Technology

This stage is the core of the research, which will be carried out through two parts. First of all, on the basis of the previous research, try to build a "block chain + university network ecological governance" work model. The basic structure of the model is "1 + 2 + 4," in which "1" is a center, which refers to the block chain technology, which is the core technology, in line with the basic purpose of this project to try to innovate the direction of computer application; "2" is two categories, which refers to the University + network ecology. With the advent of the Internet era, the network has become an important way for college teachers and students to obtain information, which provides a good carrier and communication platform for ideological work in colleges and universities. The complexity, virtuality, timeliness, openness and other characteristics of the Internet also bring severe challenges to ideological work in colleges and universities. The team members of the research group are all from colleges and universities. As educators, they must firmly hold the ideological position of colleges and universities, cooperate with party committees at all levels to jointly build an orderly and clean university network ecosystem; "4" is the four dimensions, including hardware environment, software environment, individual subject, and institutional subject. Secondly, this topic will focus on "1 + 2 + 4" in-depth research, using block chain data structure to verify and store the data of each link of network ecology, explore the introduction of block chain corresponding technology, establish an open Internet sharing and safe enough architecture, and then realize the distributed data storage and management system of university network ecosystem with data communication, business communication, clear responsibility and authority, and reliable data. It can effectively promote the reasonable integration of block chain technology and university network ecological governance, and finally form a relatively complete working model of "block chain + university network ecological governance."

2.5 Practice of University Network Ecological Governance Model Based on Block Chain Technology

This stage will first select suitable universities to carry out research work. The standard of practice is moderate scale of school running, the coverage of subjects, categories, sources of teachers and students, and registered residence of teachers and students are wide. The sample chosen for practice is universal and has good reference value. Secondly, the typical links and prominent problems of network ecological governance in colleges and universities will be screened. From the aspect of network hardware, it will focus on the practice of network security; from the aspect of network software, it will focus on the optimization practice of the current representative network system and data platform of colleges and universities; from the aspect of teachers and students, it will focus on the early warning and supervision of network public opinion, network behavior norms, and the construction of integrity system. We should carry out the practice in this aspect. Finally, through a series of targeted practices, we summarize the successful experience, put forward reasonable suggestions for the network ecological governance of colleges and universities, and point out the direction for further research.

3 Outlook

The research results will be applicable to solve the common problems of network ecological governance in most colleges and universities, and also provide innovative ideas for ideological work, student management, and ideological and political education in colleges and universities, so as to effectively expand and enrich the existing work mode. To solve the important needs of the governance and improvement of the network ecosystem faced by colleges and universities, further solidly promote the political task of consolidating the ideological position of colleges and universities, and carry out the management of colleges and universities innovatively, so as to make a certain contribution to the construction goal of improving the level of running a school, effectively serving the local regional economy, and transporting talents with high quality, and create a clean and upright network for the whole society cultural environment.

References

1. Chen, L.D., Yi, Z.L.: Key Words of Communication. Beijing Normal University Press, Beijing (2009)
2. Jing, J.H., Han, K.L.: The construction of public opinion ecology in university network. Heilongjiang Higher Education Research, 127–129 (2015)
3. Huang, J.L.: Empirical research on network ecology and mainstream ideology communication in Colleges and universities. Party Construction and Ideological Education in Schools, pp. 42–45 (2016)
4. Du, P.J.: Research on the construction of University Network Ideological and political education ecosystem under the environment of media convergence. School Party Construction and Ideological Education, pp. 50–51 (2018)
5. Li, L.L.: Formation and guidance path of university network opinion leaders in micro era. Educ. Theory Pract. 31–33 (2020)
6. Satoshi nakamoto. Bitcoin: A peer-to-peer electronic cash system, https://bitcoin.org/bitcoin.pdf, last accessed 2017/11/23
7. Huang, J.F.: Review of block chain technology. J. Beijing Univ. Posts Telecommun. 1–8 (2018)
8. Dong, M.: Research on the architecture of University intelligent financial system based on block chain technology. Friends Account. 157–161 (2020)

Research on Informatization Construction of University Laboratory Under the Background of New Infrastructure

Peng Duan and Ying Li

Abstract The construction of 5G network, data center and other new infrastructure, marks the opening of new infrastructure in China, which is also an inevitable measure to serve the transformation of new and old kinetic energy and inject new vitality into economic development in the information age. How to implement the new infrastructure solidly and effectively and how to promote the high-quality development of the new infrastructure are the key research directions in the next step. This article will take the university laboratory information construction as the breakthrough point, to explore how the new infrastructure technology reasonably integrates into the existing mode, how to innovate the existing mode, and try to provide a reasonable reference for improving the university laboratory information construction level and constructing the diversified management mode.

1 Proposal of New Infrastructure Concepts

The world economy has entered the new pattern of international fierce competition characterized by innovation driven. The major developed countries have stepped up the development of strategic emerging industries in an attempt to seize the commanding heights of technological innovation and industrial development in the future. China should vigorously increase the proportion of new technology in the industrial structure and use new technology to promote the development of Internet, big data, artificial intelligence, manufacturing, and other real economy [1].

At the central economic work conference in December 2018, the decision makers stressed that we should play a key role in investment, increase the technological transformation and equipment renewal of manufacturing industry, speed up the pace of 5Gbusiness, and strengthen the construction of new infrastructure such as artificial intelligence, industrial Internet, and Internet of things. New infrastructure construction, as a new term, began to appear in national-level documents [2].

P. Duan · Y. Li (✉)
Shandong Technology and Business University, Yantai, China
e-mail: 202013815@sdtbu.edu.cn

© The Author(s), under exclusive license to Springer Nature Singapore Pte Ltd. 2021　　171
L. C. Jain et al. (eds.), *3D Imaging Technologies—Multidimensional Signal Processing and Deep Learning*, Smart Innovation, Systems and Technologies 236,
https://doi.org/10.1007/978-981-16-3180-1_22

In March 2020, the Standing Committee of the Political Bureau of the CPC Central Committee held a meeting to speed up the construction of 5G network, data center, and other new infrastructure [3]. The new infrastructure has become the focus of the society and industry.

2 Current Situation of Laboratory Information Construction in Colleges and Universities

With the rapid development of the information age, the informatization construction level of university laboratory not only directly affects the comprehensive management quality of the laboratory, but also affects the teaching quality, scientific research level, and school running efficiency [4].

The management of college laboratories has always been a complex project involving people, finances, and things. Judging from the current management model, most daily management tasks are cumbersome and trivial, which will consume a lot of manpower and material resources, but will still show the characteristics of low work efficiency and high error rate. In recent years, various colleges and universities have carried out a certain degree of information construction in response to the school situation, which has effectively alleviated related contradictions, but there are still many common problems that need to be resolved urgently.

2.1 There are Many Business Systems, and the Information of Materials and Equipment Cannot Be Shared

The informatization construction process of many universities is gradually advancing. Therefore, during the construction process, different bidding vendors and different business needs will inevitably have multiple business systems coexisting. Each department are fighting each other, and it is difficult to form data sharing. As a result, it is more difficult to query information and data statistics, and the degree of sharing of laboratory materials and equipment is not high.

2.2 The Investment in Information Construction is Insufficient

Informatization construction is inseparable from the strong support of funds. The low level of informatization will lead to insufficient funding for additional equipment management, severely affect the laboratory environment, and have a significant impact on the overall teaching and research environment of the college or department [5]. The second manifestation is the lack of comprehensive planning. On the one hand, the lack of information about existing materials and equipment before purchasing equipment has led to the failure to achieve the most effective purchase decisions. On the other hand, many university laboratories have a relatively shallow understanding of the scope of informatization construction. They only stop at introducing some microcomputer equipment, electrical equipment, multimedia equipment, adding some security cameras, adding several sets of business management systems, and other infrastructure constructions, leading to informatization. Most of the construction still stays at the level of traditional manual management, and the level of information construction is not high.

2.3 The User Interaction Experience is not Strong

The existing information construction of university laboratory has made great efforts in the management of equipment, goods, and personnel, but the lack of information interconnection among the three parties, the lack of personalized services for laboratory personnel or students, the lack of the use of interactive tools and interactive software, and the teaching experiment activities cannot achieve the optimal effect.

2.4 The Informatization Level of Managers is Uneven

Different experimental teaching centers have different requirements for the informatization level of management staff, which leads to the inconsistent process of informatization promotion. At the same time, the service level of relevant managers also shows uneven state.

3 Innovation Path of University Laboratory Information Construction

3.1 The Angle of Laboratory Construction

Under the background of new infrastructure construction, college laboratories can innovate ideas, break the original model, and consider rebuilding laboratories from three levels: the Internet of IOT layer, the interconnection layer, and the application layer.

Among them, the establishment of the Internet of Things layer requires the establishment of RFID identification equipment and various wired and wireless sensor equipment based on actual needs on the basis of traditional hardware to collect and store information and data such as various materials, equipment, and laboratory environment.

The construction of the interconnection layer needs to rely on the characteristics of the 5G cellular ultra-dense heterogeneous network, a reasonable layout, and build the interconnection and intercommunication of multi-modal networks, especially the seamless coverage of wireless networks, to ensure the stability of data interaction at the IOT layer; application layer identification. The terminal supports diversification and convenience, and data collection and processing are timely and accurate.

Finally, the construction of the application layer requires the penetration of artificial intelligence interactive tools, through certain rules setting, to build a unified management platform, complete the data interaction and sharing at the IOT layer and the interconnection layer, so as to provide a safe, stable, efficient, rich, and convenient experiment service.

3.2 Laboratory Management

The construction of laboratory management in colleges and universities focuses on the two aspects of material equipment and personnel management. First, let us talk about the innovative management ideas of laboratory materials and equipment. Previously, each laboratory basically cooperated with schools or secondary colleges to establish a certain scale of asset management registration system. When materials and equipment were purchased and put into storage, they were set up with electronic numbers, and then manually enter the asset system, and later use the laboratory's teacher and student feedback to maintain, repair, or scrap the equipment according to the wear and tear of the equipment. This process seems to have accurate statistics on materials and equipment, but the disadvantages are obvious. Laboratory managers are passive in the management of equipment. They need to check themselves or use feedback to collect information, which will inevitably cause information acquisition delays or errors, resulting in a waste of resources.

The M2M (machine-to-machine, M2M) technology in 5G technology can solve this drawback well. The definition of M2M mainly has two kinds of broad sense and narrow sense. In a broad sense, M2M mainly refers to machine-to-machine, human-to-machine, and mobile network-to-machine communication. It covers all technologies that realize communication between humans, machines, and systems. In a narrow sense, M2M only refers to machine-to-machine communication. Communication between machines, intelligent and interactive are typical characteristics that distinguish M2M from other applications. The machines under this characteristic are also given more "wisdom" [6].

How to make materials and equipment become smart? When they are in place, use the RFID identification system to create electronic tags for them through specific reading and writing devices, and upload relevant attribute information to the big data cloud platform simultaneously, so that each device A material is given a unique electronic identity like a person. In addition, a suitable high-frequency RFID identification system is constructed at the entrance and exit of the laboratory to automatically complete the warehouse access and location information update management of materials and equipment. At the same time, the attribute information of materials and equipment recorded on the big data cloud platform will also be fed back to the manager through various terminals such as mobile terminals to provide detailed information on maintenance, repair, and scrap in accordance with pre-established rules, as well as feedback on the use effect evaluation data within a certain period of time and provide rich decision-making reference for managers, and no manual intervention is needed in the later stage of the whole process. In addition, according to actual needs, different laboratory materials and equipment data of different departments can be integrated. The personnel who set the authority can call and view the required information by themselves, or apply to the laboratory management department, which can greatly improve the utilization of the laboratory effectiveness.

Next, let us talk about the innovative management ideas of laboratory personnel. The laboratory personnel are equipped with working magnetic cards, and establish electronic identity information for the cards through the read–write device. According to the management authority, the identity card has functions such as smart access control switch, personnel, material and equipment location positioning, material and equipment information viewing, and laboratory environment monitoring. At the same time, the corresponding application software is installed through mobile phones and other mobile terminals, and it has multiple humanized functions such as 5G network high-definition calls, online multi-person collaboration, one-key switch laboratory appliances, security alarm tracking, and personnel leave.

3.3 Laboratory Application

The application level of college laboratories focuses on the aspects of laboratory security measures, laboratory environment construction, laboratory educational affairs arrangements, and laboratory resource acquisition.

First of all, in terms of fire protection, with the electronic tag attributes assigned to the material and equipment at the laboratory management level, by matching the RFID identification system and infrared sensor equipment to the camera, the administrator can monitor the attribute information of the material and equipment through a mobile or PC terminal, especially fire hazards such as abnormal temperature in order to prevent fires in advance. For theft prevention, the laboratory can add a face recognition sensor, which can track and compare the personnel entering and exiting the laboratory. If there is theft of materials and equipment, it can cooperate with the public security department in the first time to lock suspects through big data analysis, which is effective in avoiding property damage. At the same time, for the safekeeping of special materials and equipment, according to needs, corresponding wireless sensing equipment such as humidity sensors can be constructed for effective protection.

Secondly, the daily management of the laboratory environment has been a headache for managers for a long time. For example, large computer laboratories often cause serious waste of resources because the computer is not turned off, the electrical power is not turned off, and the air conditioner is not turned off. With the technical support of the new infrastructure, in the future, through the implementation of similar solutions for smart homes, various necessary sensor devices will be constructed and transmitted to the cloud server through the 5G network. After intelligent analysis, the control requirements will be automatically fed back to the relevant equipment, and the environment will be realized. The automatic adjustment is turned off.

In terms of the arrangement of laboratory educational administration, there is a common phenomenon in the laboratories of colleges and universities. Because the individual subject direction or research group considers the urgent need of the equipment when purchasing the laboratory equipment, the equipment will be used in one or several time periods, and the equipment will be idle for a long time in most of the remaining time Due to the waste of teaching resources, relying on the new technology brought by the new infrastructure, we can consider opening laboratory resources in an all-round way. In addition to the more reasonable course scheduling by the educational administration department, teachers and students can also visit the big data cloud platform of the laboratory at any time and apply for extracurricular use through the mobile terminal. In this way, we can plan all laboratory resources as a whole, realize the maximum embodiment of the value of experimental resources.

In terms of laboratory resource acquisition, through the interactive system, each class operation of the laboratory can be used to form electronic information. At the same time, teachers can also selectively record or broadcast the classroom teaching process through recording and broadcasting equipment. The teaching resources above the teacher side can be used as a necessary reference to improve the teaching effect, and the students can choose to call teaching resources through various terminals after class to consolidate relevant knowledge. In addition, related majors and even related universities can build a larger-scale big data cloud platform to realize seamless connection and sharing of experimental teaching data to facilitate mutual exchange and learning.

4   Summary and Prospect

This article revolves around the conceptual connotation of the new infrastructure, mainly around the two aspects of 5G technology and data center construction, and tries to make innovative suggestions for the construction of university laboratory informatization. Many suggestions are only at the stage of theoretical conception, but the author believes with the further support of relevant policies for the new infrastructure and the promotion of the position of decision makers in college laboratories, the new infrastructure is bound to bring a qualitative leap in the informatization construction of college laboratories.

References

1. Xi, J.P.: Xi Jinping on State Administration, Vol. 3. Foreign Languages Publishing House Co., Ltd., Beijing (2020)
2. The central government set the tone, "new infrastructure" is completely on fire! These seven science and technology fields will explode. https://tech.sina.com.cn/roll/2020-03-05/dociimxxs tf6692640.shtml, last accessed 2020/03/05
3. What new kinetic energy is contained in the construction of new infrastructure?, http://www.xin huanet.com/2020-03/06/c_1125673676.htm, last accessed 2020/03/06
4. Luo, L.M.: Practice and research on the construction of laboratory informatization in private colleges under the background of internet. Comput. Knowl. Technol., p. 162–1649 (2017)
5. Li, Q.: Exploration and practice of the open model of professional laboratories. Exp. Technol. Manage., pp. 224–226 (2017)
6. Zhao, G.F.: Overview of 5G mobile communication network key technologies. J. Chongqing Univ. Posts Telecommun. (Natural Science Edition), p. 444 (2015)

Run-Time Prediction Practices of Multimedia Web Design in Technology Management

Bin Hu, Sohail M. Noman, Muhammad Irshad, Xilang Tang, Chen Song, and Musa Uba Muhammad

Abstract Modern multimedia platform design becomes more complex. In a single device, a multimedia system (MMS) requires the support of many applications. Increasingly, processors are integrated to achieve high peak performance in systems. Due to the application domain and digital hardware, MMS heterogeneity has also increased since time algorithms were supported. To achieve system performance, limited resources must be shared. Interference occurs when applications run dynamically across shared resources. Why don't you run these applications? Because adding dimensions to the design problems. Most companies spend 65–75% of the total cost of product development on the verification of use cases. Only analytical techniques can reduce development costs. Analysis in product design is not feasible as applications are added to the system during runtime. Therefore, the article includes a run-time practice for multimedia design's performance that can predict the performance of multiple applications before they run accurately and quickly in the system.

B. Hu
Changsha Normal University, Changsha, China

S. M. Noman (✉)
Shantou University Medical College, Shantou, Guangdong, China

M. Irshad
Department of Electronic and Information Engineering, The Hong Kong Polytechnic University, Hung Hom, Hong Kong

X. Tang
Airforce Engineering University, Xi'an, China

C. Song
Tianjin University of Science and Technology, Tianjin, China

M. U. Muhammad
Kano University of Science and Technology, Wudil, Nigeria

© The Author(s), under exclusive license to Springer Nature Singapore Pte Ltd. 2021   179
L. C. Jain et al. (eds.), *3D Imaging Technologies—Multidimensional Signal Processing and Deep Learning*, Smart Innovation, Systems and Technologies 236,
https://doi.org/10.1007/978-981-16-3180-1_23

1 Introduction

By the time Internet came, all the communication industries get very fast progress [1–3]. Today, any user from any corner of the world can share and access multimedia documents through the internet [4, 5]. It is possible that in near future each user will be capable of network computing. The system is capable of processing multimedia applications and data; such a system is known as a multimedia system. In recent times, a modern trend has been introduced of multimedia Web-based applications. According to different research teams, multimedia Web-based applications are now a very important tool to get information on the Internet [6–8]. The main reason that is their platform is independent, interactive, and accessible. The Web-based applications become characterized into two types, allowing the user to access the database and other is to allow end users to perform analysis, and multimedia Web-based systems become characterized by the storage, processing, performance, transmission, and representations of several times-dependent and time-independent media streams.

Today's technology is growing faster. Many new requirements are coming on the way to design Web [6, 9, 10]. The question arises in mind that how multimedia Web-based design system helps customers or fulfill their needs? The system must be designed in such ways that it helps users to buy stuff easily, to find information, to save time and money, and to help them by online conversations in case of any difficulty or query. Reduce the problem ratios as much as possible, such as in the period of technology management, some multimedia Web sites talk about sales and some do services either in business or online solutions. Multimedia Web-based projects can help to determine the structure of sites by clarifying the main objectives and purpose.

1.1 Multimedia System Characteristics and Trends

The involved characteristics in this system include: the system must be integrated, and all the information handled must be digitally represented. Besides, the interface must be interactive to the final presentation of media [11, 12]. In integrated, and, the world's first console video game released by Magnavox supported so many games from tennis to baseball [13]. To interconnect the different logic and signal generators, removable circuit cards (consist of series of jumpers.) became used to produce the required output components (logic and screen). When it was released, it came with some plastic (translucent) overlays because when it was put on to the television screen, it generates color images but, at that time, the sound was not supportable by it. That is the reason, it becomes known as the first-generation console video game. Now move forward to the present time, where seventh generation is around of the video games consoles [14, 15]. The best example is the "Play station" that came into the market by Sony. Besides sounds and colors, this invention also supports the whole media center, which is capable to play movies in high definition, photographs, and also video games in the advanced format, and is designed with such large sizes of hard disks to store games, movies, and other kinds of stuff. In adding features, it can also be connected from the home network (wired and wireless) to the whole world. From the past till the present, we have come up with the development of multimedia systems. From both the application and system design prospects, a lot of evolution has been made. Besides everything, many key challenges remain.

Further, MMS uses a direct combination of content forms (text, audio, video, pictures, and animation) to entertain users by providing different information [16]. The gaming console is one of the multimedia systems examples. Besides this console, there present so many other examples like television, home theater, mobile system technologies, mp3 players, and laptops. Technology nowadays becomes so advanced, users become to expect of getting information online whether while traveling on a plane or sitting in a café. A large number of multimedia products are available on the Internet to fulfill the user demands, and the reason is that features are gradually increasing day by day, for example, a mobile phone that is designed for voice calls now become supporting the video-conferencing features and 5G network technology is now using for streaming television programs [17]. Another with mp3 players, which were designed for just playing music but now it turned to additional features like the store, play photographs, video clips, and video games also. Many applications are being executed with different combinations on the platform. This thing is sometimes called the scenario in literature [11–13, 16]. Let us take the example of the mobile phone again because at one instant while surfing on the Internet and downloading the applications in the background of doing call at the front and in the other instance, a mobile phone can be used to listening to music while playing with the photographs stored in the phone and on meanwhile, it can be used to do remote devices to access the file manager in the phone. The multimedia technologies like health care [18–20] and power consumptions [21] become the basic feature in designing. The following trends come to summarize the multimedia system applications as [22–24]:

1. Many applications are coming in a multimedia system.
2. The number of use cases (executable applications in combinations) becomes gradually increasing.
3. The market pricing of the products is reducing because of a lot of competition and interfaces.

1.2 Challenges in Web-Based Multimedia Systems

The above two sections indicate the complexity gradually increasing in modern multimedia design. With performance and resources, they can support the number of running applications. When the designer is creating the design of the system, some challenges come to his mind like reduce cost and the main one is to design it on time. To reduce the cost, the designer has to explore the number of design options and keeps in mind to search out the closest solution. Also, the performance of the running applications in the system evaluates the user skills. By summarizing, the challenges which come on the way while designing the Web-based multimedia system includes: multiple use-cases, designing and programming, design techniques, run-time applications, and performance [25–29].

Analysis: Performance plays a vital role. The accurate calculation of the time waiting is the main point to perform analysis. When an application is designed for a model in the form of graphs, their performance can easily be calculated. It could be computed when it is executing. At the same time when it runs, it depends on the scheduler (whether it is static or dynamic), it finds out that whether the settlement of resources is fixed by designing time or these are chosen at running time. In earlier cases, the graphs present the execution order and performance of the system. This thing is mostly working for static applications, and a dynamic scheduler is required for the dynamic applications (of multimedia). Further, some key points for analysis methodology include: multimedia systems are fast enough, it can handle a very large number of applications, and it also allows the additional applications to come.

Design: Designers use to face a lot of challenges to reduce the marketing time in designing and testing system for use-cases. Swift prototyping is very important to explore and evaluate the alternatives (hardware and software). But most work is done by hand because of fewer techniques and tools which make the design process time consuming by keeping errors. However, some efforts and techniques were made by designers to make a routine flow of work but these are still limited to designing a single application. Multimedia system not only supports multiple applications but also supports the multiple use-cases. The high demand for functionalities in the modern multimedia system is going toward the development (programmable hardware and software) to increase the flexibility of design.

Management: The resources present in the system and managing them is known as resource management. Which includes the resource allocator and the starting up of applications? As much the multimedia design is concerned, the major difference from a general system is: the applications which become usable by domains are generally known. The analyzing of the design time becomes harder because of reasons like some understand about the applications at designing time that can be used in the future, and sometimes at design time, the accurate platform may not be known [30, 31]. The fundamental components involve in multimedia system design are shown in (Table 1).

2   Elements of Web-Based MMS

Multimedia elements play a vital role in Web design, and it is all about creativity. To make the Web site more captive, innovative, and more interactive, a Web designer can add multimedia elements professionally like text, audio, graphics, sound effects, and animation with the help of different programming tools like silver light [32, 33].

Animation: To make the Web site more interactive for the users, Web designers put animations on pages. The best example is GIF format in new gaming Web sites which grab the attestation of the audience in one look. Not only gaming but there

Table 1 Key components of MMSD

Components	Example
Capture devices	Video camera, recorders, audio, graphics, 3D
Storage devices	Hard disks, CD-ROM, zip drives, etc.
Communication networks	Ethernet, Internets, intranets, ATM
Computer systems	Multimedia machines, MPEG hardware
Display devices	HDTV, Hi-Res monitors and color printers, etc.

so many other sites are present which play the best example role for multimedia elements. In cartoons, multimedia does perfect work.

Graphics/Videos: Because of graphics and video interfaces, many like invitations come by users. This reason is based on Web site popularity among users from worldwide. Users are tired to read the long PDF files. Multimedia elements make it easier for users to understand any topic with the help of videos; "adobe flash player" on YouTube is the best example in the present time. Videos are only the way to get traffic on Web sites.

Audios: On the one hand, graphics/videos have their impact, and, on the other side, audios also playing a business environmental role. Audio with picture impacts on user's mind, and the user gets attracted by clearly understanding messages verbally. Its best example comes in educational multimedia Web sites, where students can understand topics easily and make their points clear with help of audio/video clips. Though the result comes impressively by using multimedia elements, it slows down the processing speed, which put an effect on site ranking by less visitor ratio.

3 Multimedia Web Design Principles

Some of the multimedia Web design principles includes [14–16, 21, 22, 25, 27, 28, 32]:

Contrast: In multimedia, contrast means to fill colors. It produces visual effects on the Web site and gets the interest of users. Without contrast stuff on the Web page is unorganized.

Proximity: With proximity, you can increase the overall organizational design. It became used for some group-related information.

Alignment: It comes to align the elements used in the multimedia design system. It creates visual unity.

Balance: If the elements are not balanced on the Web page, it creates tension for users and diverts the attention. By using the different weights of phrases, it looks messy for the user and makes it hard for the user to understand the thing quickly.

Repetition: It is something similar to consistency, by repeating the visual elements again and again.

Layout: While designing a multimedia Web, themes, and layout, the basic principle do not place a challenge for the user to find out the things.

4   Use of Techniques of Web-Based MMS

The techniques in the development of multimedia are written from the literature. These techniques include traditional structure, use of different case modern techniques, and storyboarding. Also, various other techniques include project management, prototyping, flowcharts, Manu maps, data flow diagrams (DFD), object oriented, relationship management data models (RMDM) diagrams, movie authoring and designs (MAD), class diagrams, entity–relationship diagrams (ERD), dialogue charts, state transition diagrams (STD), functional decomposition diagrams (FDD), and use-case diagrams (UCD) [30, 31, 34].

An interesting fact of MMS techniques usage like MAD, which is widely used techniques in is of MMS Web design. It became a shred of evidence that project management and multimedia prototype are openly used techniques. Flowchart and blog diagrams techniques are used in navigational structure pages. In Web design, the most used programming languages are visual basic and C + + , SQL, and Java also. In the development process, it is hard to understand the object-oriented techniques used [4, 5, 35].

5   Conclusion

Multimedia system technology has always had a heavy side benefit. It produces efficiency and also helps to reduce the price. In addition, this run-time prediction practice analysis shows the multimedia components and elements that can be incorporated in Web design in technology management, by increasing the effects elements in the structure of the multimedia Web design, but the processing may slow down. Today, the small–medium enterprises are not ready to turn to multimedia aspects, but soon, the majority of organizations will become part of multimedia design in technology management by adopting new effects and upgrading hardware/software systems.

Acknowledgements The initial objective draft of this article was introduced by one of the authors at the education meeting in Stamford college, Malaysia, and getting considered to be a part of this proceedings later on.

References

1. Qiao, X.: Integration model for multimedia education resource based on internet of things. Int. J. Continuing Eng. Educ. Life Long Learn. **31**(1), 17–35 (2021)
2. Qureshi, A., Megías Jiménez, D.: Blockchain-based multimedia content protection: review and open challenges. Appl. Sci. **11**(1), 1 (2021)
3. Sugar, W., Hoard, B., Brown, A., Daniels, L.: Identifying multimedia production competencies and skills of instructional design and technology professionals: an analysis of recent job postings. J. Educ. Technol. Syst. **40**(3), 227–249 (2012)
4. Sadiq, M.W., Hameed, J., Abdullah, M.I., Noman, S.M.: Service innovations in social media & blogging websites: enhancing customer's psychological engagement towards online environment friendly products. Revista Argentina de Clínica Psicológica **29**(4), 677–696 (2020)
5. Huo, C., Hameed, J., Haq, I.U., Noman, S.M., Chohan, S.R.: The impact of artificial and non-artificial intelligence on production and operation of new products-an emerging market analysis of technological advancements a managerial perspective. Revista Argentina de Clínica Psicológica **29**(5), 69–82 (2020)
6. Kumar, S. N., Fred, A. L., Padmanabhan, P., Gulyas, B., Dyson, C., Kani, R. M., Kumar, H. A.: Multimedia-based learning tools and its scope, applications for virtual learning environment. In Computational Intelligence in Digital Pedagogy, Vol. 197, pp. 47–63, Springer, Singapore (2021)
7. Shah, S.M.A., Ge, H.W., Haider, S.A., Irshad, M., Noman, S.M., Arshad, J., Ahmad, A., Younas, T.: A quantum spatial graph convolutional network for text classification. Comput. Syst. Sci. Eng. **36**(2), 369–382 (2021)
8. Sohail, M.N., Jiadong, R., Muhammad, M.U., Chauhdary, S.T., Arshad, J., Verghese, A.J.: An accurate clinical implication assessment for diabetes mellitus prevalence based on a study from Nigeria. Processes **7**(5), 289–307 (2019)
9. Yu, J.: Data sensing based inventive system for multimedia and sensing of web applications. In 2020 Fourth International Conference on Inventive Systems and Control (ICISC), pp. 437–440, IEEE (2020)
10. Ionescu, B., Müller, H., Péteri, R., Abacha, A. B., Datla, V., Hasan, S. A., Demner-Fushman, D., Kozlovski, S., Liauchuk, V., Cid, Y. D., Kovalev, V., Pelka, O., Friedrich, C. M., Seco De Herrera, A. G., Ninh, V. T., Le, T. K., Zhou, L. T., Piras, L., Riegler, M., Halvorsen, P., Tran, M. T., Lux, M., Gurrin, C., Dang-Nguyen, D. T., Chamberlain, J., Clark, A., Campello, A., Fichou, D., Berari, R., Bire, P., Dogariu, M., Ştefan, L. D., Constantin, M. G.: Overview of the ImageCLEF 2020: multimedia retrieval in medical, lifelogging, nature, and internet applications. In International Conference of the Cross-Language Evaluation Forum for European Languages, pp. 311–341, Springer, Cham (2020)
11. Sing, A. L. L., Ibrahim, A. A. A., Weng, N. G., Hamzah, M., Yung, W. C.: Design and development of multimedia and multi-marker detection techniques in interactive augmented reality colouring book. In Computational Science and Technology, Vol. 603, pp. 605–616, Springer, Singapore (2020)
12. Castro-Medina, F., Rodríguez-Mazahua, L., López-Chau, A., Cervantes, J., Alor-Hernández, G., Machorro-Cano, I.: Application of dynamic fragmentation methods in multimedia databases: a review. Entropy **22**(12), 1352 (2020)
13. Copenhaver, A., Griffin III, O. H.: White-collar criminality within the video game industry. Games Culture, 1555412020975629 (2020)
14. Gill, T., Ma, Z., Zhao, P., Chen, Y. K.: How accessories add value to a platform: the role of innovativeness and nonalignability. Euro. J. Market. (2020)
15. Nieborg, D. B.: Apps of empire: global capitalism and the app economy. Games Culture, 1555412020937826 (2020)
16. Sabet, S. S., Schmidt, S., Zadtootaghaj, S., Naderi, B., Griwodz, C., Möller, S.: A latency compensation technique based on game characteristics to mitigate the influence of delay on cloud gaming quality of experience. In Proceedings of the 11th ACM Multimedia Systems Conference, pp. 15–25, ACM (2020)

17. Dang, S., Amin, O., Shihada, B., Alouini, M.S.: What should 6G be? Nat. Electron. **3**(1), 20–29 (2020)
18. Sohail, M.N., Jiadong, R., Irshad, M., Uba, M.M., Abir, S.I.: Data mining techniques for medical growth: a contribution of researcher reviews. Int. J. Comput. Sci. Netw. Secur **18**, 5–10 (2018)
19. Sohail, N., Jiadong, R., Uba, M., Irshad, M., Khan, A.: Classification and cost benefit analysis of diabetes mellitus dominance. Int. J. Comput. Sci. Netw. Secur **18**, 29–35 (2018)
20. Sohail, M.N., Ren, J.D., Uba, M.M., Irshad, M.I., Musavir, B., Abir, S.I., Anthony, J.V.: Why only data mining? a pilot study on inadequacy and domination of data mining technology. Int. J. Recent Sci. Res **9**(10), 29066–29073 (2018)
21. Liu, W., Lombardi, F., Shulte, M.: A retrospective and prospective view of approximate computing [point of view. Proc. IEEE **108**(3), 394–399 (2020)
22. Lacuška, M., Peráček, T.: Trends in global telecommunication fraud and its impact on business. In Developments in Information & Knowledge Management for Business Applications, pp. 459–485, Springer, Cham (2021)
23. Chakraborty, S.: Active learning for multimedia computing: survey, recent trends and applications. In Proceedings of the 28th ACM International Conference on Multimedia, pp. 4785–4786, ACM (2020)
24. Sohail, M.N., Jiadong, R., Uba, M.M., Irshad, M., Bilal, M., Akbar, U., Rizwan, T.: Forecast regression analysis for diabetes growth: an inclusive data mining approach. Int. J. Adv. Res. Comput. Eng. Technol. (IJARCET) **7**(9), 715–721 (2018)
25. Husić, J. B., Baraković, S., Krejcar, O., Maresova, P.: Modeling of quality of experience for web-based unified communications with perceptual dimensions. Signal, Image Video Process. pp. 1–9 (2021)
26. Mladenović, M., Ošmjanski, V., Stanković, S.V.: Cyber-aggression, cyberbullying, and cyber-grooming: a survey and research challenges. ACM Comput. Surveys (CSUR) **54**(1), 1–42 (2021)
27. Al-Garadi, M.A., Varathan, K.D., Ravana, S.D.: Cybercrime detection in online communications: the experimental case of cyberbullying detection in the Twitter network. Comput. Hum. Behav. **63**, 433–443 (2016)
28. Liu, P., Guberman, J., Hemphill, L., Culotta, A.: Forecasting the presence and intensity of hostility on Instagram using linguistic and social features. In Proceedings of the International AAAI Conference on Web and Social Media, **12**(1), (2018)
29. Singh, V.K., André, E., Boll, S., Hildebrandt, M., Shamma, D.A.: Legal and ethical challenges in multimedia research. IEEE Multimedia **27**(2), 46–54 (2020)
30. Kumar, A. (2009). Analysis, design and management of multimedia multiprocessor systems. Ph. D. dissertation. Technische Universiteit Eindhoven, (2009)
31. Kuma,r A., Corporaal, H., Mesman, B., Ha, Y.: Multimedia Multiprocessor Systems. Dordrecht: Springer Netherlands, (2010)
32. Hidayati, A., Bentri, A., Yeni, F.: The development of instructional multimedia based on science, environment, technology, and society (SETS). J. Phys.: Conf. Series, **1594**(1), 012016, IOP Publishing (2020)
33. Cetto, A., Netter, M., Pernul, G., Richthammer, C., Riesner, M., Roth, C., Sänger, J.: Friend inspector: a serious game to enhance privacy awareness in social networks. arXiv preprint arXiv:1402.5878, ACM Comput Surv (2014)
34. Barry, C., Lang, M.: A survey of multimedia and web development techniques and methodology usage. IEEE Multimedia **8**(2), 52–60 (2001)
35. Hao, W., Shah, S.M.A., Nawaz, A., Nawazc, M.A., Noman, S.M.: The Impact of CPEC on infrastructure development, a-double mediating role of project success factors and project management. Revista Argentina de Clínica Psicológica **29**(4), 737–750 (2020)

Design and Simulation of Orthogonal Frequency-Division Multiplexing (OFDM) Signaling

Bin Hu, Muhammad Irshad, Sohail M. Noman, Xilang Tang, Chen Song, and Sami Ahmed Haider

Abstract The orthogonal frequency-division multiplexing (OFDM) is a modulation scheme which is spectrally efficient and which decomposes one carrier into several subcarriers. Each subcarrier is orthogonal, allowing the transfer of data on each subcarrier simultaneously. This work performs the strengths of OFDM and quadratic amplitude modulation (QAM) in parallel on a communication channel. A MATLAB program was written to examine and investigate the communication systems of the OFDM by comparing OFDM and QAM single carrier. This technique is significant for the complex simulates systems. The demonstration of two graphical user interfaces (GUI) incorporates the simulation concepts of OFDM.

1 Introduction

The focus of this study is on multiplexing and simulation in orthogonal frequency-division multiplexing (OFDM) by resistance resulting to intersymbol interference (ISI) [1–3], because channelizing of the OFDM suits high-speed communication.

B. Hu
Changsha Normal University, Changsha, China

M. Irshad (✉)
Department of Electronic and Information Engineering, The Hong Kong Polytechnic University, Hung Hom, Hong Kong SAR
e-mail: mirsha@polyu.edu.hk

S. M. Noman
Shantou University Medical College, Shantou, Guangdong, China

X. Tang
Airforce Engineering University, Xi'an, China

C. Song
Tanjin University of Science and Technology, Tanjin, China

S. A. Haider
Department of Computing, University of Worcester, Worcester, UK

© The Author(s), under exclusive license to Springer Nature Singapore Pte Ltd. 2021
L. C. Jain et al. (eds.), *3D Imaging Technologies—Multidimensional Signal Processing and Deep Learning*, Smart Innovation, Systems and Technologies 236,
https://doi.org/10.1007/978-981-16-3180-1_24

The time frame of every transmission reduces as communications systems improve their information transfer speed. Since multipath delays remain unchanged in high data communications, ISI is considered as a limitation [4]. In addition, this article analyzes and examines the projective theory validation of effective coded OFDM and GNU Radio implementation given by Bineta Sarr et al. [5] and Duc Toan et al. [6], also, the wireless speech and video transmission given by Onsy Alim et al. [7]. OFDM avoids this problematic issue through simultaneous transmission of many low-speed transmissions, in example of the four pieces of binary data as shown in (Fig. 1) in two ways [8–11]. Assume it takes four seconds for this transmission. Then every data piece in the left image is about one second long.

On the contrary, as shown in the right, OFDM sends four pieces. Each data part is four seconds long in this case. This extended period leads to fewer ISI problems. OFDM's other consideration is to implement high-speed systems in a low-complexity fashion compared to conventional single-train systems [12–15].

One of the fundamental technical decisions in architecture of all communication systems is the choice of a modeling system [16–19]. The criteria on which the different modulations are to be determined must be defined before modulation is chosen. Bandwidth efficiency is also the key element. The subsequent multipath propagation of the diffuse link also limits the possible bandwidth of the channel. This also results in a primary metric of bandwidth efficiency [20]. The third element is the reliability of the transmission. In adverse conditions, a modulation technique must also provide a minimum acceptable error rate and should show multipath resistance to ISI and variation in components of the DC signal.

It has been noticed in the researches, that, in high-speed communication, ISI has always been considered as the significant problem [21–24]. It occurs when there is an interference in the transmission and the recipient becomes unable to decode the transmission properly. Such as, the same transmission is sent in every direction in a wireless communication system (WCS) like the one as presented in (Fig. 2). More than a copy of signal reaches to the recipient, given that, signal reflects larger objects such as building or a mountain. This is called multipath in communication terminology. Various ICT devices (information and communication), like mobiles, iPad, etc., do contain multiple communication interfaces such as Ethernet, Wi-Fi, and HSDPA/LTE [9–11, 17, 25]; but, for technical reasons, one of them can only

Fig. 1 Traditional versus OFDM communication

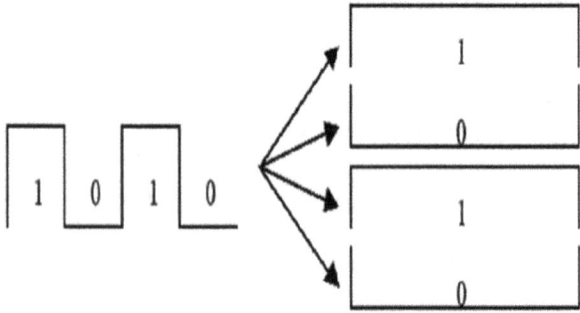

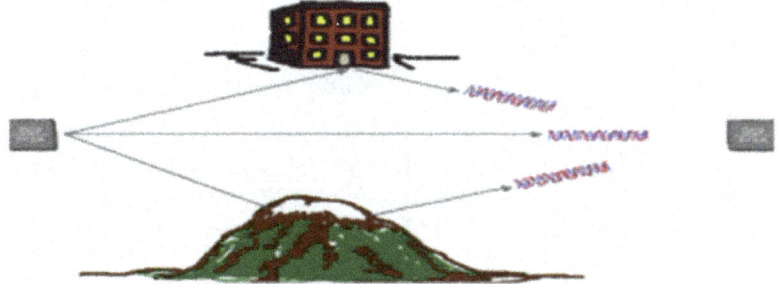

Fig. 2 Multipath demonstration

be used at once: The IP address plus the port number are the endpoint of TCP/IP communication, and the IP is always connected to the network interfaces. Late copies of signal interfere with the direct signal and result in ISI, given that indirect path takes more time for the receiver to go.

2 Objective

This study holds and provides two project goals including: (1) QAM simulation single carrier; and (2) simulation of multicarrier OFDM and then comparison of QAM single carrier and OFDM multicarrier to show multipath OFDM channels strength.

3 Methodology

3.1 Background

The main technique for coping with variable wireless channels is diversity in wireless systems such as (802.11). Diversity is the spread of information through various independently faded channels with a certain redundancy. If this is complete, there is a little probability of successful communication being prevented from deeply fading on one single channel [13, 26, 27]. However, the trick is to find faded channels independently. This comes from the time, frequencies, and wireless link spatial resources. These exist in the physical layer. The focus of this project will be on multiplexing and simulation in the orthogonal frequency-division multiplexing (OFDM). Because of its ISI resistance, OFDM considered suitable for the communication of high speed. Each transmission time reduces as communications systems improve their information transfer speed. Since multipath delays remain constant, in contrast, ISI becomes a constraint in communication at high data rates. By simultaneously sending various

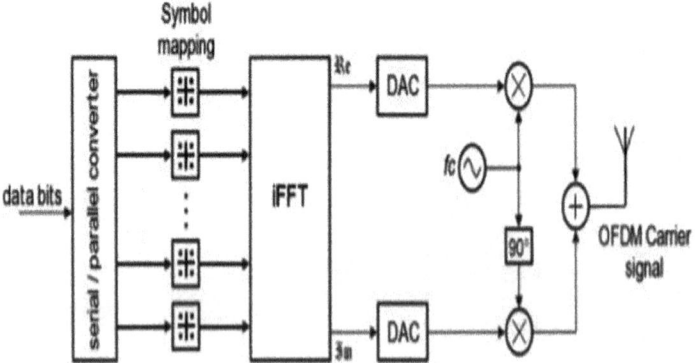

Fig. 3 OFDM demodulation diagram

transmissions of low speed, OFDM avoids this problem. The use of OFDM and multicarrier modulation has been highlighted over the last few years as it is an ideal platform for the transmission of wireless data as illustrated in (Fig. 3).

The block diagram of OFDM demodulation can be achieved digitally using an inverse fast Fourier transform (IFFT). The OFDM demodulation is virtually the inverse operation of the modulation. As OFDM is highly sensitive to frequency and phase noise, pilot symbols are added to the data packet. They are used to estimate the channel transfer function, and the inverse is applied to every subcarrier OFDM signal to compensate for the channel, so that OFDM can simplify the equalization process by turning the frequency-selective channel into a flat channel. During demodulation, the pilot subcarriers that were known by the receiver are used to correct the phase and amplitude of the received OFDM signals.

However, during the research on how interference can be reduced between close-spaced canals, the concept of OFDM technology was first explored during the 1960s and 1970s [4, 28–31]. Furthermore, in the face of interference and selective propagation conditions, it was necessary for error-free transfer of data. In addition, the use of OFDM initially required high processing levels and was therefore not viable for overall use.

Furthermore, digital broadcasting was one of the primary systems to adopt OFDM [32–35]. OFDM has been able to offer a highly reliable form of data transport under several signals. The DAB digital radio was once an example and introduced in Europe and elsewhere. Norwegian Broadcasting Corporation (NRK) has carried out the launch of first service by June 1, 1995 [36]. Digital TV was also used with OFDM. This paper summarizes the main contributions as follows:

(1) The implementation of a new block of RC code and convolutionary encoding on the GNU Radio to carry out a forward error correction code.

(2) Installation design and laboratory-controlled noise measurements: It is a novel technique for using GNU Radio rather than an oscilloscope as proposed and suggested for other laboratory tests.

(3)   Over the laboratory impulsive noise air experiment: Several scenarios are developed in light of the impulse level, the distance between emitter, recipient, and source; and line-of-sight (LOS) and non-line-of-sight (NLOS) situations are included.

Further, the article is organized as the overview of the system is presented in (Sect. 2). Firstly, before describing the transmitter and recipient designed on the GNU Radio, it shows the simulation chain. In a controlled laboratory, the setup for impulsive noise measurements is detailed in (Sect. 3) with certain findings. The results achieved indicate the impulsive noise can be accurately reproduced by this novel strategy. The specifications and measuring scenarios are provided in (Sect. 4). The results obtained will be presented in (Sect. 5).

3.2 Preprocessing

Initially, the performance of network has been detailed by considering the traffic rate with average delays, which resulted in the efficiency of LOS in comparison of N-LOS. After that, bit error rate (BER) performance was considered in the presence of impulsive noise. The results were satisfactorily even if a little degradation can be observed in some cases. This section is also overviewed for a comparison between the simulation results demonstrated in [20], and the resulted findings, confirming that several impulsive noise voltages maintain the same performance. Finally, the conclusion is given with perspectives of (Fig. 4).

GNU Radio platform holds the software modules and setup, which are important in process modulation and demodulation of OFDM. For instance, an application of OFDM has been offered in GNU Radio's directory, which includes files like (benchmark_tx.py), (benchmark_rx.py), and optional (enchmark_add_channel.py). In this context, GNU Radio-based implementation of OFDM systems is developing. It can be regarded as the simple simulated example of OFDM system that includes transmitter and receiver created by the system and software. Signals are transmitted via

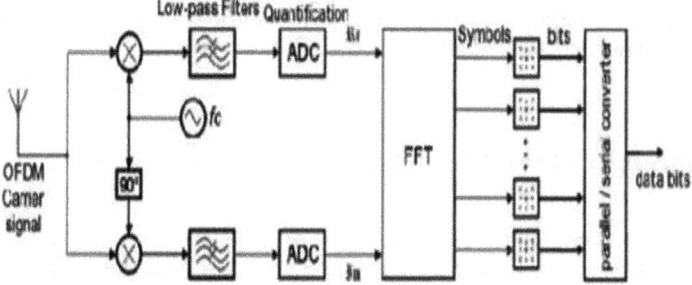

Fig. 4 Implementation of OFDM technique with GNU Radio

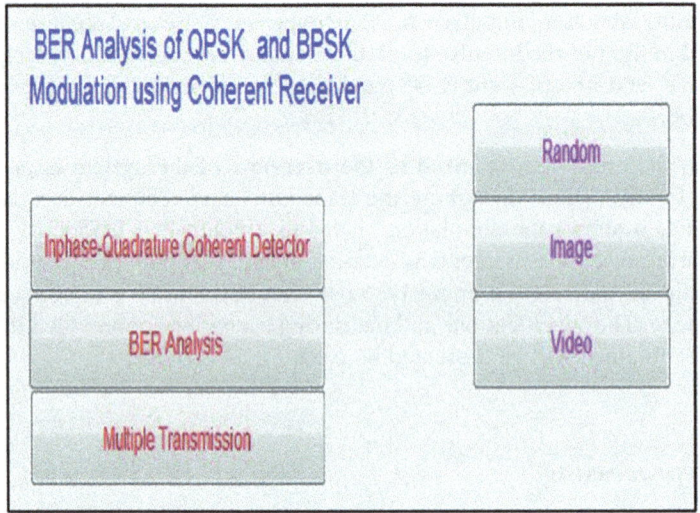

Fig. 5 MATLAB GNU Radio source program

USRP devices via realistic environments. Software can also run the OFDM system completely. In this case, benchmark add channel.py will simulate the entire transmission chain including the channel. This example was used as the basic source code for the OFDM communication system in this project. Later power in the processing of the 4G mobile communications system, which began its implementation from around 2009, increased because of increased integration levels [37]. OFDM is now considered. For Wi-Fi and various other wireless data systems, OFDM was also adopted. The OFDM's basic principle is to separate data in N lower rates at high rates from a data source.

GNU Radio software (GNU stands for GNU Not Unix) is opened and the most common software and hardware used in SDR systems (USRP). Each individual can build, modify, and develop his system as shown in (Fig. 5) because of its open-source platform. Further, the flow is modulated by PSK or QAM. Simultaneously, these streams are transmitted by a summary to a symbol OFDM on orthogonal subparts. Mathematically if X [n] n = 0, 1,..., N−1 is the complex input of subpopulation n, N is the subcarrier number and T is the symbol duration.

3.3   Flowchart of the Methodology

This research and simulation project involves an OFDM communication system. The MATLAB simulation code is shown in (Fig. 6) as a simplified flowchart.

First, the transmitter converts input data to parallel sets from a serial stream. For every subcarrier, each dataset contains one symbol, S. For instance, a set would be

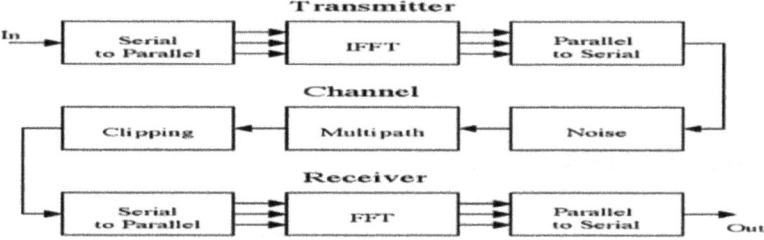

Fig. 6 OFDM simulation flowchart

[S0 S1 S2 S3] with four data. This example set as in (Fig. 7) before performing IFFT is arranged on the frequency-domain horizontal axis.

To use IFFT to manipulate these data, this symmetric arrangement on the vertical axis is required. Fourier's reverse transformation transforms the frequency-domain dataset into samples of these data's time-domain representation. The IFFT is especially useful to OFDM as it generates frequency compatible samples with orthogonal conditions. The OFDM signal is created parallel to the serial block and the time-domain samples are out sequences.

The simulation of the channel allows the testing of common features, such as noise, multipaths, and clipping. Simple noise is simulated by adding random data information to the transmitted signal. Multipath simulation involves copying the transmitted signal in the original with attenuation and delay. This simulates the problem when the signal spreads across many wireless paths, for example, through a direct path or path that rebounds a building a receiver can view a signal. Clipping also simulates the amplifier problem saturation. This addresses the practical implementation problem in OFDM, where the peak-to-average power ratio is great. The receiver reverses the transmitter. Initially, the OFDM data is divided into parallel serial stream sets. The fast Fourier transform (FFT) returns an image of a frequency domain with the time-domain sample. The magnitude of the frequency components is based on their

Fig. 7 Frequency-domain distribution of symbols

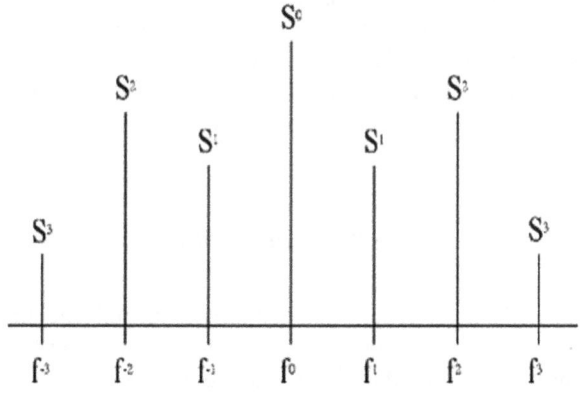

original data. Finally, the parallel serial block transforms this parallel data into a serial data recovery stream.

4 Results and Discussion

The MATLAB simulation accepts text and audio inputs as well as binary, sinusoidal, or random information. The corresponding OFDM is then generated, a channel simulated, the input data recovered, and the error rate is analyzed. In order to compare OFDM with a conventional single-carrier communications system, a 16-QAM simulation can be carried out. These simulations are dynamic and allow the user to set parameters that define the characteristics of the communication system. A graphical user interface was used for two simple demonstrations of OFDM communication following the MATLAB toolbox type demonstrations (image and video). This allows researchers and scientists to quickly learn the basic concepts of OFDM communication.

The first demonstration is to create an OFDM symbol with basic GUI (or basic GUI_win). A simple example is the use of Fourier's transformation to send binary data to four frequencies. First, there are certain irregularities in the curves. This can be explained by the critical aspect of the error performance of the RC limit distance decoder. A boundary distance, since the received word is received, either procures a word or confirmed a failure within the agreed range of the given word. In the first case, the decoding error occurred when the generated code text did not correspond to the codeword sent. Furthermore, the challenges and subject of research is identified as a wireless communication on-board. Several common commercial protocols, such as digital video broadcast (DVB), asymmetric digital subscriber line (ADSL), and wireless Ethernet (Wi-Fi), implement OFDM. With Wi-Fi, the IEEE 802.11a and IEEE 802.11 g implementations specifically use OFDM techniques. With IEEE 802.11 g, each channel occupies 16.25 MHz of bandwidth at the 2.4 GHz frequency range. In addition, each channel is divided into 52 subcarriers of 312.5 kHz. Together, these subcarriers overlap to fully utilize the 16.25 MHz channel bandwidth dedicated per channel. In addition, each subcarrier can use a unique modulation scheme. More specifically, Wi-Fi can use BPS, QPSK, 16-QAM, or 64-QAM depending on the characteristics of the physical channel being used. One of the newest wireless Internet protocols, WiMAX, also uses OFDM technology.

4.1 Abbreviations and Acronyms

OFDM: Orthogonal frequency-division multiplexing: A single data stream splits across different narrow-band channels at different frequencies, which is a method of digital signal modulation to reduce interference and crosstalk.

5 Conclusion

This MATLAB simulation demonstrates that OFDM is more suitable for multitrack transmission than a single-carrier transmission technology like 16-QAM. The system performance is assessed with multiple measurements, such as the packet delivery ratio, average time, and the BER. The results obtained are satisfactory globally and confirm the claim. It may be interesting in the perspective of other tests to increase further the distance between the emitter and the source and the recipient. Just a few studies relate to impulsive noise reduction, as indicated in introduction part of studies. It would also be interesting to implement the proposed method in future work for this purpose, which will enable our experimental results to be compared.

References

1. Long, S., Khalighi, M.A., Wolf, M., Bourennane, S., Ghassemlooy, Z.: Investigating channel frequency selectivity in indoor visible-light communication systems. IET Optoelectron. **10**(3), 80–88 (2016)
2. Komine, T., Haruyama, S., Nakagawa, M.: A study of shadowing on indoor visible-light wireless communication utilizing plural white LED lightings. Wireless Pers. Commun. **34**, 211–225 (2005)
3. Xu, X.M., Ni, L., Zhang, J.Y., Liu, L.J., Deng, L.W., Cao, C., Du, Z.: Inter-symbol interference and equivalent voltage noise of high-speed serial system. J. ZheJiang Univ. (Eng. Sci.). **48**(1), 118–123 (2014)
4. Keller, T., Hanzo, L.: Adaptive multicarrier modulation: a convenient framework for time-frequency processing in wireless communications. Proc. IEEE **88**(5), 611–640 (2000)
5. Sarr, N.B., Agba, B.L., Gagnon, F., Boeglen, H., Vauzelle, R.: Analysis and experimental validation of efficient coded OFDM for an impulsive noise environment. Sensors **18**(11), 3667 (2018)
6. Nguyen, D. T.: Implementation of OFDM systems using GNU Radio and USRP. University of Wollongong, (2013)
7. Alim, O. A., Shaaban, S., Bahy, H.: Speech and video transmission over wireless ATM networks. In Proceedings of the Twenty Third National Radio Science Conference (NRSC'2006), pp. 1–7, IEEE (2006)
8. Raymo, F.M.: Digital processing and communication with molecular switches. Adv. Mater. **14**(6), 401–414 (2002)
9. Sadiq, M.W., Hameed, J., Abdullah, M.I., Noman, S.M.: Service innovations in social media & blogging websites: enhancing customer's psychological engagement towards online environment friendly products. Revista Argentina de Clínica Psicológica **29**(4), 677–697 (2020)
10. Shah, S.M.A., Ge, H.W., Haider, S.A., Irshad, M., Noman, S.M., Arshad, J., Ahmad, A., Younas, T.: A quantum spatial graph convolutional network for text classification. Comput. Syst. Sci. Eng. **36**(2), 369–382 (2021)
11. Huo, C., Hameed, J., Haq, I.U., Noman, S.M., Chohan, S.R.: The impact of artificial and non-artificial intelligence on production and operation of new products-an emerging market analysis of technological advancements a managerial perspective. Revista Argentina de Clínica Psicológica **29**(5), 69–82 (2020)
12. Wang, X.Y., Gao, L.J., Mao, S.W., Pandey, S.: CSI-based fingerprinting for indoor localization: a deep learning approach. IEEE Trans. Veh. Technol. **66**(1), 763–776 (2016)
13. Adame, T., Bel, A., Bellalta, B., Barcelo, J., Oliver, M. IEEE 802.11 ah: the WiFi approach for M2M communications. IEEE Wireless Commun. **21**(6), 144–152 (2014)

14. Popoola, O.R., Sinanović, S., Popoola, W.O., Ramirez-Iniguez, R.: Optical boundaries for LED-based indoor positioning system. Computation **7**(1), 7 (2019)
15. Lencse, G., Kovács, Á.: Testing the channel aggregation capability of the MPT multipath communication library. In Proc. World Symposium on Computer Networks and Information Security, pp. 13–15 (2014)
16. Irshad, M., Liu, W., Wang, L., Shah, S. B. H., Sohail, M. N., Uba, M. M.: Li-local: green communication modulations for indoor localization. In Proceedings of the 2nd International Conference on Future Networks and Distributed Systems, pp. 1–6, ACM (2018)
17. Hao, W., Shah, S.M.A., Nawaz, A., Nawazc, M.A., Noman, S.M.: The impact of CPEC on infrastructure development, a-double mediating role of project success factors and project management. Revista Argentina de Clínica Psicológica **29**(4), 737–750 (2020)
18. Sohail, M.N., Jiadong, R., Irshad, M., Uba, M.M., Abir, S.I.: Data mining techniques for medical growth: a contribution of researcher reviews. Int. J. Comput. Sci. Netw. Secur **18**, 5–10 (2018)
19. Sohail, M.N., Ren, J.D., Uba, M.M., Irshad, M.I., Musavir, B., Abir, S.I., Anthony, J.V.: Why only data mining? a pilot study on inadequacy and domination of data mining technology. Int. J. Recent Sci. Res **9**(10), 29066–29073 (2018)
20. Fiandrino, C., Kliazovich, D., Bouvry, P., Zomaya, A.Y.: Performance and energy efficiency metrics for communication systems of cloud computing data centers. IEEE Trans. Cloud Comput. **5**(4), 738–750 (2017)
21. Song, X., Zhao, Z., Wei, Y., Wang, M.: Research and performance analysis on multi-point light-emitting diodes-array intra-vehicle visible light communication systems with DC-biased optical–orthogonal frequency-division multiplexing. Int. J. Distrib. Sens. Netw. **13**(9), 1550147717729785 (2017)
22. Armstrong, J., Schmidt, B.J.: Comparison of asymmetrically clipped optical OFDM and DC-biased optical OFDM in AWGN. IEEE Commun. Lett. **12**(5), 343–345 (2008)
23. Jiang, T., Tang, M., Lin, R., Feng, Z., Chen, X., Deng, L., Fu, S.N., Li, X., Liu, W., Liu, D.: Investigation of DC-biased optical OFDM with precoding matrix for visible light communications: theory, simulations, and experiments. IEEE Photonics J. **10**(5), 1–16 (2018)
24. Zhang, X., Zhou, Y. F., Yu, Y. P., Han, P. C., Wang, X. R.: Comparison and analysis of DCO-OFDM, ACO-OFDM and ADO-OFDM in IM/DD systems. In Applied Mechanics and Materials, Vol. 701, pp. 1059–1062, Trans Tech Publications Ltd (2015)
25. Khan, M.M.A., Shahriar, N., Ahmed, R., Boutaba, R.: Multi-path link embedding for survivability in virtual networks. IEEE Trans. Netw. Serv. Manage. **13**(2), 253–266 (2015)
26. Mahmood, A., Exel, R., Trsek, H., Sauter, T.: Clock synchronization over IEEE 802.11—A survey of methodologies and protocols. IEEE Trans. Ind. Informatics, **13**(2), 907–922 (2017)
27. Adame, T., Bel, A., Bellalta, B., Barcelo, J., Oliver, M.: IEEE 802.11 ah: the WiFi approach for M2M communications. IEEE Wireless Communications, **21**(6), 144–152 (2014)
28. Azim, A.W., Le Guennec, Y., Maury, G.: Hermitian symmetry free optical-single-carrier frequency division multiple access for visible light communication. Optics Communications **415**, 177–185 (2018)
29. Lohrasbipeydeh, H., Gulliver, T.A.: Unknown RSSD-based localization CRLB analysis with semidefinite programming. IEEE Trans. Commun. **67**(5), 3791–3805 (2019)
30. Pathak, P.H., Feng, X., Hu, P., Mohapatra, P.: Visible light communication, networking, and sensing: A survey, potential and challenges. IEEE Commun. Surv. Tutorials **17**(4), 2047–2077 (2015)
31. Nguyen, H.Q., Choi, J.H., Kang, T.G., Lim, S.K., Kim, D.H., Kang, M., Lee, C.G.: Effect of LED emission cross-section in indoor visible light communication systems. EURASIP J. Wirel. Commun. Netw. **2012**(1), 1–11 (2012)
32. Al-Moliki, Y.M., Alresheedi, M.T., Al-Harthi, Y.: Physical-layer security against known/chosen plaintext attacks for OFDM-based VLC system. IEEE Commun. Lett. **21**(12), 2606–2609 (2017)
33. Aminikashani, M., Gu, W., Kavehrad, M.: Indoor positioning with OFDM visible light communications. In 2016 13th IEEE annual consumer communications and networking conference (CCNC), pp. 505–510, IEEE (2016)

34. Jiang, W., Wright, W. M.: Indoor wireless communication using airborne ultrasound and OFDM methods. In 2016 IEEE International Ultrasonics Symposium (IUS), pp. 1–4, IEEE (2016)
35. Ghimire, B., Seitz, J., Mutschler, C.: Indoor positioning using OFDM-based visible light communication system. In 2018 International Conference on Indoor Positioning and Indoor Navigation (IPIN), pp. 1–8, IEEE (2018)
36. Krumsvik, A. H.: Strategy and structure for online news production–Case studies of CNN and NRK. In International Handbook of Internet Research, pp. 325–340, Springer, Dordrecht (2009)
37. Khan, A. H., Qadeer, M. A., Ansari, J. A., Waheed, S.: 4G as a next generation wireless network. In 2009 International conference on future computer and communication, pp. 334–338, IEEE (2009)

Research on Intelligent Financial Framework of Colleges and Universities in Information Age

Yelin Zhong, Caihong He, and Jiaqi Tang

Abstract This paper expounds the connotation and characteristics of intelligent finance, analyzes the key technology of intelligent finance, and introduces an intelligent financial framework model for universities and colleges, which at least includes infrastructure service layer, data service layer, software service layer, interaction service layer, as well as corresponding information standard system and security technical guarantee system. At last, an application context for intelligent finance is provided.

1 Introduction

With the rapid development of information technology, the financial system of all works of life will undergo a revolutionary change. It can be said that every time in human history, the development of information technology will have a profound influences on the financial area. For instance, the emergence of the computer replaces traditional manual bookkeeping; the emergence of internet helps finance realize telecommuting. The organizational mode and financial process of the enterprise have undergone significant changes. The shared financial service center arises with the time, which promotes the integration of businesses and finance and the development of enterprise process reengineering. With the integration and mixture of mobile Internet, big data, Internet of things, artificial intelligence, cloud computing, chain blocks, and 5G technology, plus the promotion of various cloud applications, like financial robots, financial cloud, E-invoice cloud, and the cloud of taxation, intelligent digital finance will completely subvert the traditional financial pattern [1].

Y. Zhong · C. He · J. Tang (✉)
School of Accounting, Hunan University of Technology and Business, Changsha Hunan 410205, China
e-mail: hutbtjq@163.com

© The Author(s), under exclusive license to Springer Nature Singapore Pte Ltd. 2021 199
L. C. Jain et al. (eds.), *3D Imaging Technologies—Multidimensional Signal Processing and Deep Learning*, Smart Innovation, Systems and Technologies 236,
https://doi.org/10.1007/978-981-16-3180-1_25

2 The Connotation and Characteristics of Intelligent Finance

2.1 Definition of Intelligent Finance

There is no unified definition for intelligent finance. We believe that intelligent finance is a kind of new financial management mode, which simulates, extends, and expands traditional finance through the cooperation between human beings and machines by means of the new technology of "big data, artificial intelligence, mobile network, cloud computing, Internet of things, block chains" represented by artificial intelligence. This mode is guided by advanced financial management theories, tools, and methods, and simplifies complicated financial management activities, in order to reduce the cost of accounting work, improve the effect of accounting work, improve the quality of accounting information, and enhance accounting compliance and value creation ability, to better the college management and support decision-making services [2, 3].

2.2 Features of Intelligent Finance

We believe that intelligent finance has the following characteristics [4, 5].

Comprehensive Sharing. It includes the sharing of financial accounting platform, data platform, personnel, and organization platform related to intelligent finance by universities and colleges. General financial shared service center focuses on accounting, while based on which the comprehensive sharing of intelligent finance will further expand the sharing from financial accounting to business and management accounting.

Integration and Synergy. On the basis of the unification of mechanism, process, system, data and standard, the deep integration and collaboration of business, finance, and management in colleges and universities would be realized.

Man–Machine Integration. Through the coordination of people and intelligent machines, the man–machine integrated intelligent system is formed to realize the complex business and financial management work. Thus, the degree of automation of business and financial management would be continuously improved to provide support for personalized decision-making services.

3 Key Technologies of Intelligent Finance [1]

3.1 Cloud Computing

The concept of "cloud computing" was first formally introduced by Eric Schmidt, Google's chief executive, at the Search Engine Conference in 2006. In the same year Amazon launched EC2, an elastic computing cloud service.

In 2012, China Academy of Information and Communications Technology proposed in *the White Paper on the Development of Cloud Computing* that cloud computing is an information processing method that realizes large-scale computing through unified organization and flexible application of various ICT information resources through the network. The decentralized ICT information resources are centralized through the Internet to form a shared resource pool, which dynamically provides services in a measurable way according to the needs of users. Users can access ICT resource services through any network connection on a variety of terminals (including but not limited to personal computers, tablets, smart phones, and even smart TVs).

That same year, the national institute of standards and technology (NIST) proposed that cloud computing is a kind of shared pool where users can easily visit computing resources (e.g., networks, servers, storage, applications, and services) on demand. Because it can quickly demand for these configurable resources, the management cost and service suppliers' intervention would be minimized.

In 2014, the ISO/IEC standardization organization released Cloud Computing-Overview and Vocabulary, which argues that cloud computing is a service model that provides network access to a pool of scalable, resilient, shared physical, or virtual resources, where users can purchase on demand through self-help.

Cloud computing has a profound impact on finance, including: (1) breaking physical space constraints and saving financial management costs. (2) Expand financial information sources and share financial information in real time. (3) Reorganize financial processes and improve data quality. (4) The integration of industry and finance makes telecommuting possible.

3.2 Big Data

In 1998, John Marcy, an American scientist, believed that with the rapid growth of data, data would be difficult to understand, obtain, process, and organize, or even worse, which triggered the thinking in the field of computing. Gartner, an authoritative research institution on big data, points out that big data is a kind of information assets with massive capacity, high growth rate, and diversification, which cannot be captured, managed, and processed within a certain time frame by conventional tools and software. New data processing modes are required.

The McKinsey Global Institute defines big data as a data set that is so large that its acquisition, storage, management, and analysis greatly exceed the capabilities of traditional database software tools. It is characterized by massive data scale, rapid data flow, diverse data types, and low value density.

The definition of International Data Corporation (IDC) is: big data is used to describe and define the massive data produced by the information society, the new architecture and technology designed, and the technology development and innovation, in order to more economically and effectively obtain the relevant value from the data with high frequency, large capacity, different types, and structures.

Inside the enterprise, the financial department, as the key node connecting various business departments, has to face more complex, massive, and changeable data. The traditional financial management mode is urgently needed to be reformed.

3.3 Artificial Intelligence

Artificial intelligence (AI), by studying the thinking mode of human brains and summarizing the thinking rules of human beings, enables computers to imitate the thinking mode of human brains through deep learning, accomplish part of the functions of human brains, and can replace human brains to solve some specific problems. Deep learning, as a kind of machine learning, establishes a neural network that simulates human brain for learning and analysis through pattern analysis to realize machine intelligence. Its basic feature is that it attempts to simulate the patterns of transmission and processing of information between neurons of the human brain. Artificial intelligence reduces the financial work intensity, improves the accuracy of financial work, and greatly enhances the intelligence level of financial work.

3.4 Mobile Internet

Mobile Internet is another new technology in the computer field. It accesses the Internet through wireless access devices and realizes the data exchange between mobile terminals. It is one of the most important directions for the future development of the network.

It can be concluded that the mobile Internet is the foundation for digital finance to realize the application of "everyone's finance."

3.5 The Internet of Things

Professor Ashton of MIT first proposed the concept of the Internet of things in 1999 when he was studying radio frequency identification (RFID) technology. He pointed out that the Internet of things is a network that connects all objects with the Internet through sensing devices such as radio frequency identification (RFID), to realize intelligent identification and management. The international telecommunication union (ITU) issued "the ITU Internet report 2005: the Internet of things" in 2005, which says the Internet of things combines the existing Internet, communication network, broadcast, and television network, and a variety of access network and private network through radio frequency identification device, infrared sensors, global positioning system (GPS), and laser scanner devices, to form a large and new network, realizing intelligent recognition and management.

4 The Basic Framework of Intelligent Finance

The construction of intelligent finance should adhere to the principles of interconnection, data sharing, application fusion, safety and reliability, and convenient expansion under the unified overall framework of intelligent campus, so as to maximize the utilization of resources and serve the business needs of university teaching, scientific research, and management to the greatest extent.

A complete basic framework of intelligent finance at least includes infrastructure service layer, data service layer, software service layer, interaction service layer, and the corresponding information standard system, and security technical guarantee system [6–8].

4.1 Infrastructure Layer IaaS

The infrastructure layer mainly includes campus network (wired network, wireless network), private network, Internet of things, sensing equipment, and other facilities. All kinds of equipment are connected and integrated for unified management. Therefore, by connecting users with users, users with objects, and objects with objects, users can easily and quickly use the network, thus realizing the interconnection of various networks and the real-time transmission of all perceptual information.

4.2 PaaS of Data Platform Layer

Data platform layer PaaS is a support platform for all kinds of data resources and applications, including data collection, storage, management, and calculation, as well as data analysis and processing application support platform, such as basic database, decision database, and resource database. Taking financial data as the core, the financial management is connected with teaching management, scientific management, and other management systems, so this platform serves as the data sharing center of "industry financial integration." Through the exchange and sharing mechanism, it provides data storage services, access to resources, and calculation for the application of the upper layer, to effectively avoid isolated data and information, providing support for the application of the upper layer.

4.3 Software Service Layer SaaS

This layer is the logical processing layer for systematic business, which provides powerful application service for colleges and universities information, and can be divided into two large application platforms: business and finance. Business application platform mainly includes the teaching systems, scientific research systems, human resource systems, student systems, asset management systems, and logistics management systems, while financial application platform mainly includes budget management, accounting management, capital management, salary management, charge management, final accounts management, project management, etc.

Based on the integration and development of business process, management process, and financial accounting process, it promotes the complete unification of business and financial data, so as to be integrated. Compared with the traditional accounting business process, financial business integration features the following characteristics: reengineering theory innovation workflow by using workflow business process, building a management system of budget, accounting, charges, salaries, assets, scientific research, and teaching centered on cloud data sharing; automatically transforming the data in business department into accounting language for the financial department in real time; generating financial data in time; embedding various management measures, internal control systems, and business flows into business software, to achieve centralized authorization and unified certification; implementing management accounting, budget accounting, new government accounting, internal control, performance evaluation, and other concepts and systems in the "integration of business and finance" service platform, so that the data of financial department and business department can be automatically transmitted in real time and the financial data can be generated in time.

4.4 Application Interaction Layer (Comprehensive Information Service, Unified Information Portal)

This layer provides users with unified access to the portal terminal. Intelligent finance integrates different applications and services and is already a unified user-oriented system when it is used by users. Different systems are also interrelated and shared. Departments are no longer islands of information. However, the traditional software and services run independently, which seriously restricts the application and construction level of the system, and causes a large amount of waste of resources. When facing with different users, application services in intelligent finance provide a variety of intelligent services in accordance with the needs of users. Meanwhile, users can log in a unified portal to query financial management information, so as to improve the efficiency of management and service quality. In the intelligent finance construction process, all systems are managed uniformly and data is shared, so as to achieve the maximized utilization of resources.

5 Application Context

5.1 Network Reimbursement System

Expense reimbursement is a widely used application context for intelligent finance [9–12].

Filling reimbursement forms. The applicant logs into the intelligent financial system, selects the items to be reimbursed, and fills in the cause and selects the item information. The applicant shall paste the original documents on the sticky sheet as required, scan them, and upload them to the financial system.

The intelligent financial system starts the image recognition engine, automatically identifies, classifies, and distributes all kinds of invoices and documents submitted by the applicant, which are finally formed into structured data.

Intelligent audit. After identifying and forming structured data, the reimbursement information is audited intelligently. Expense reimbursement audit rules are set manually and embedded into the expense reimbursement system. The system starts the rule engine and performs audit operations according to the set logic, such as checking and rechecking the authenticity of invoices, analyzing the reasonableness of amounts, and evaluating the necessity of expenses. Finally, the system will generate the audit results: First, if the data is correct, the system will automatically generate the reimbursement voucher and pass the information to the next link. Second, if the error is not mandatory, the applicant shall be reminded to give explanations, and the explanation information shall be passed to the examination and approval link. Third, if the error is mandatory, the error information will be fed back to the reporter directly, and the reimbursement of the problem documents will be rejected.

Electronic approval. The system starts the process engine and delivers the approved information to the approver step by step. The system will also remind the approver to pay attention to the non-mandatory errors detected. The examiner can view the structured data and the information of the original documents on the system. If there is no error in the audit, the system will automatically generate the reimbursement voucher. The applicant can print the voucher and send it to the financial department together with the original voucher.

Automatic payment. After the approval of the reimbursement form, the payment form will be automatically generated. The system will automatically execute the payment operation according to the payment plan and wait for the subsequent manual approval of the payment application.

Accounting treatment and records. Automatically generate vouchers and journal records according to bookkeeping rules.

In the context of networked reimbursement, the standardization and electronics of recording processes provide opportunities for the automation of accounting validation. The system advances the accounting information system from the preparation of accounting vouchers to the business process and moves the focus of the accounting information system from accounting vouchers to original vouchers. The system generates accounting vouchers according to the established rules, and accountants only need to maintain the corresponding classification and summary mechanism, to realize the accounting information systemization from data collection to data output, which greatly reduces the basic workload of accountants.

6 Conclusions

With the continuous development of colleges and universities, higher requirements have been put forward for the construction of financial management and financial system, so higher education institutions should accelerate the construction of financial information, help financial management integrate into the business, and enhance the efficiency of financial management staff, thereby raising the level of financial management, achieving sound and rapid development of colleges and universities.

Intelligent finance is still a new field. Therefore, it is hoped that continuous exploration in this field will fully promote the development of intelligent finance.

Acknowledgements A project supported by Scientific Research Fund of Hunan Provincial Education Department (17A118) (Research on Social Co-governance of Food Safety Based on Cooperative Governance Theory).

References

1. Peng, J., Chen, H., Wang, Z.X., Hu, R.Y., et al.: Digital Finance. Tsinghua University Press, Beijing (2020)
2. Liu, Q., Yang, Y.: Discussion on the architecture, implementation path and application trend of intelligent finance. China Manage. Account. Studies **1**, 84–90+96 (2020)
3. Liu, M.L., Huang, H., Tong, C.S., Liu, K.: Study on the basic framework and construction conception of intelligent finance. Account. Res. **3**, 179–192 (2003)
4. Huang, C.Y.: Study on intelligent finance features and their relationship. China Manage. Informationization **23**(19), 72–74 (2020)
5. Wang, W.Q., Xia, P.J., Hu, Y.: Study on documentary evaluation in intelligent finance. Econ. Trade Update **23**, 36–37 (2020)

6. Xie, Q.L., Cui, B., Zhang, S., Liu, T.Z. et al.: Study on top design of university financial management informationization from the perspective of smart campus. J. Qingdao Univ. Sci. Technol. (Soc. Sci.), **35**(4), 58–62 (2019)
7. Lai, J.W.: Study on constructing smart campus by internet of things and cloud computing. J. Fujian Comput. **36**(6), 44–48 (2020)
8. Yu, S.W.: Design and research on university financial information platform during the 13th "five-year" plan—smart campus construction based on "internet+ education." Commercial Account. **6**(11), 67–70 (2017)
9. Chen, H., Sun, Y.C., Guo, Y., Zhao, Y.N.: Financial robots—financial application of RPA. Finance Account. **16**, 57–62 (2019)
10. Chen, H., Zhao, Y.N., Dang, M.M.: Information practices of global financial sharing about ZTE. Finance Account. **15**, 24–26 (2015)
11. Liu, N.: Research on application of intelligent financial systems in scientific institutes. China Chief Financial Officer, **5**, 80–83 (2019)
12. Hu, T., Huang, W.S.: Study on innovative reimbursement of university finance based on bigdata technology. J. Hefei Univ. Technol. (Soc. Sci.), **32**(5), 128–133 (2018)

VHF Band Dual-Channel High-Power Transmitter and Receiver Module

Lingling Zeng and Zhongming Zhu

Abstract In this paper, a transmitter and receiver (T/R) component subsystem is designed, which is mainly composed of transmitting and receiving units. It has function of single channel taking out 2000 W peak power and double channel output 1000 W peak power, as well as the function of receiving signals. The circuit of the T/R module is designed with multilayer board, air-tight cavity and local gold plating, multilayer board, and radio frequency (RF) connector screw are installed on the cavity, low-frequency connector is installed with screw, high-temperature wire transfer is used inside, and the process is mature and reliable. It has the characteristics of high output power, large amplitude limit power, large number of channels, and high requirement of amplitude and phase consistency. At the same time, the component uses the key technology of health management and adaptability to improve the quality and reliability of the product and can perform multi-signal processing.

1 First Section

With the development of radar science in the twenty-first century, phased array radar is more and more widely used, and the T/R component is the core of active phased array antenna. According to the expansion of radar applications, in order to improve the reliability and anti-jamming of electronic equipment technology, active phased array T/R components are constantly developing on the road of miniaturization and lightness. In this paper, a VHF band with high power and low noise is designed with active phased T/R components [1].

This paper introduces a T/R two-channel assembly that can be used at high altitudes, uses 6061 aluminum alloy material to make cavities, and uses tungsten copper as a gold-plated carrier to promote the amplifier, so that the device can quickly conduct heat and reduce the coefficient of expansion. With an input power of 5 dBm, a dual-channel output power of >1000 W and a single-channel output power of

L. Zeng (✉) · Z. Zhu
School of Information Science and Technology, Chengdu University of Technology, Chengdu 610059, China

© The Author(s), under exclusive license to Springer Nature Singapore Pte Ltd. 2021
L. C. Jain et al. (eds.), *3D Imaging Technologies—Multidimensional Signal Processing and Deep Learning*, Smart Innovation, Systems and Technologies 236,
https://doi.org/10.1007/978-981-16-3180-1_26

>2000 W can be achieved [2]. And this component has a healthy adaptive technology, enhances the debugging and reliability of the component, and improves product life.

2 Scenarios and Structures

2.1 The Overall Scenario

It is known that the T/R component subsystem is applied in the terrestrial environment, and the system has high requirements on the reliability of the components. The working environment temperature ranges from minus 20 to above 50 °C, which is cooled by air-cooled method (see Fig. 1).

Due to the complexity and tension of T/R component circuit, T/R component small signal design adopts multilayer board design, while component digital circuit is designed separately, monitoring unit wave control board is independently designed to weld PCB printing board, and T/R components are divided into multi-cavity layout. The component layout consists mainly of the emission unit, the receiving unit, and the control unit. In order to achieve good shielding of components and reduce costs, components are multilayer cover, local gas sealing structure. According to the characteristics of the circuit, it is divided into five independent cavities, namely four radio frequency cavities, one digital cavity, the radio frequency (RF) cavity is used to design RF frequency transmission of RF frequency circuits, and through this segmentation to ensure channel isolation; the power supply between each cavity is connected by high-temperature wires, and the signal portion is connected by capacitors or copper skins. Due to the large number of control lines, component small signal circuit and RF transmission use multilayer board design, low-frequency signal, and

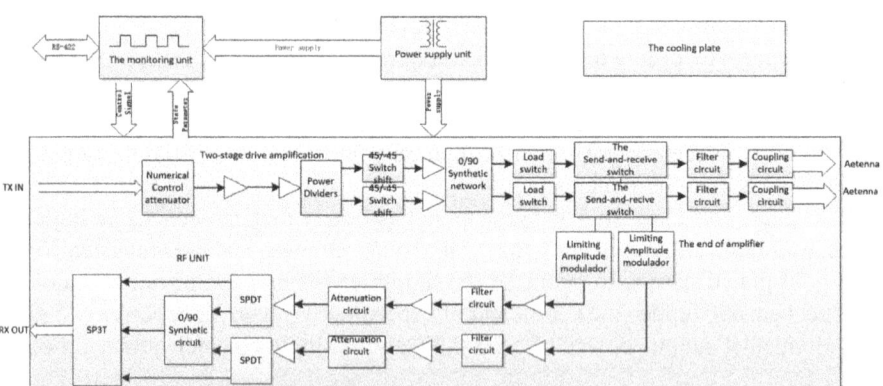

Fig. 1 Schematic diagram of T/R components

The schematic of the transmitter circuit

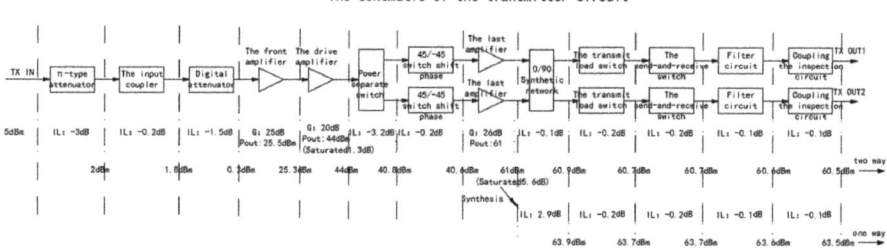

Fig. 2 Schematic diagram of amplifying link and circuit of transmitting unit

high frequency signal layered line, to ensure high and low frequency and the separation between power supply and control, and RF transmission all use compartments separate, reduce crosstalk; The power input is filtered [3].

2.2 The RF Frequency Link Design of the Transmitting Unit

Working frequency range of transmitting unit: 50 MHz ± 2.5 MHz; Input power: (5 ± 1) dBm; Output power: ≥2000 W; it mainly includes six circuit parts: voltage-controlled attenuation circuit, driving amplifier circuit, power divided phase shift circuit, final amplifier circuit, 0/90° synthetic network circuit, filter circuit, coupling detection circuit, filter circuit, transceiver switch circuit, etc. The detailed emission unit principle block diagram is shown in Fig. 2.

How it works: The RF excitation signal is first fed into the input coupler for signal detection for pulse width and aspect ratio detection, and then into the CNC attenuator to achieve multi-speed power switching, and then into the RF power amplification. RF power amplification uses the front stage, drive, and the last stage, three-stage series amplification. After the signal is amplified by the drive, it enters the merit circuit and controls the two-way end-stage amplified output single 2KW or two-way 1KW by switching phase shift and send-off switch, and the amplified signal is coupled by filter wave for output power detection and finally output to the antenna.

As shown in Fig. 2, the transmission link is driven by a front-stage push, which eventually drives the power amplification of the two-way end-stage mode, with a 25 W LDMOS amplifier tube, a gain of 20 dB, and a 1300 W LDMOS amplifier with a gain of 26 dB.

2.3 The RF Frequency Circuit Design of the Receiving Unit

The operating frequency range of the receiving unit is 50 ± 2.5 MHz, and the dynamic range of the receiving signal is ≥65 dB (−40 to −105 dBm) which is used to amplify

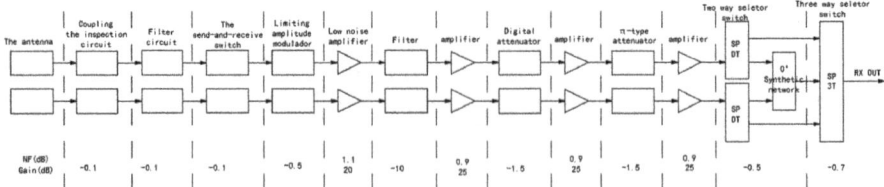

Fig. 3 Schematic diagram of amplifying link and circuit by receiving unit

the signal received by the antenna, multi-stage low-noise banished amplified signal, with limit protection function, filter function, gain control function, etc. (see Fig. 3).

The working principle: The antenna receiving signal enters the receiving channel through a coupling circuit, a filter circuit, and a radio and receiving switch, and enters a low-noise circuit after the limiter limit. The low-noise circuit consists of multi-stage low-noise amplifier, sound meter filter, CNC attenuator, etc., which mainly realizes the low-noise amplification filtering and gain control function of RF signal. The amplified signal selects the output via a switch.

The receiving unit RF link uses four-stage low noise to ensure gain greater than 80 dB, and a six-digit attenuator to ensure a gain adjustable range of ≥30 dB. The noise figure of the receiving unit is mainly determined by the insertion loss before the low-noise amplifier and the noise figure of the first-stage low-noise amplifier, so the low-insertion loss limiter and the first-stage low-noise amplifier with the low-noise figure are used to ensure the noise figure of ≤2.5 dB. In order to ensure the out-of-band suppression of the receiving channel, the requirement of out-of-band suppression is achieved by using the sound meter filter.

2.4 Control Circuit

The control circuit takes STM32F407VET6, a 32-bit ARM single-chip microcomputer, as the core control circuit. It receives the RS422 control information from the upper computer and completes the control of A/B transceiver switch, A/B phase shift switch, A/B load switch, three-option receiving switch, A/B receiving attenuation, and other information. Test the A/B standing wave and A/B power of the component; pulse width and duty ratio of input signal are detected. Check the current and temperature of the components.

(1) Hardware circuitry: The hardware circuit mainly includes interface circuit, detection circuit, output control circuit, and main control circuit.

    (A) Composed of MAX3491ESD, it completes the conversion of 422 difference level to TTL level and TTL level to 422 difference level.

    (B) The detection circuit completes the current and temperature detection of the component, the A/B standing-wave detection, the A/B power detection, and the input pulse width and duty ratio detection. If any of the parameters tested above are abnormal, the component will immediately turn off the power amplifier switch and the transceiver switch to protect the other circuits of the component.

Various voltage detection signals output by the RF circuit are sent to the voltage follower and then from the voltage follower to the A/D port of STM32F407VET6.

(A) The main control circuit is composed of 32-bit ARM SINGLE-chip MICROCOMPUTER STM32F407VET6, which analyzes the input asynchronous serial data and outputs the control code required by the RF circuit, that is, generating various switch control signals and attenuation signals, and carrying out A/D transformation for various analog signals.

(B) The output control circuit is composed of SN74ALVC164245DGGR. Various control signals generated by the single chip can be converted from +3.3 V level to +5 V level through SN74ALVC164245DGGR to drive the RF circuit.

(2) Software design: According to component design indexes, C programming language is used to realize the analysis of control protocol, output control to RF circuit and BIT detection.

3 Key Technologies

3.1 Health Management and Adaptive Techniques

In this T/R module, temperature, current, voltage, gain, and other sensing data are relatively important. The microprocessor is used as the core of the control circuit to complete intelligent perception and adaptive adjustment. Its principle block diagram is shown in Fig. 4. T/R components in the long-term working process, due to environmental temperature, power supply fluctuations, and device aging conditions, its various indicators will also change. When the fluctuation of each working index of T/R component is detected, relevant voltage and current values are collected through the ADC and sampling resistance built into the adaptive power supply and uploaded to the microprocessor [4]. The microprocessor issues the correction instruction to the adaptive power supply, which outputs the analog control voltage through the built-in DAC to adjust the output voltage of the power module in real time. When severe fluctuations cannot be corrected in time (such as overpressure and over temperature), immediately turn off the power supply and report the cause. Adopting the advantages of adaptive digital power supply which is very outstanding, it can provide real-time intelligence, can automatically adapt to the environment, and improves the efficiency

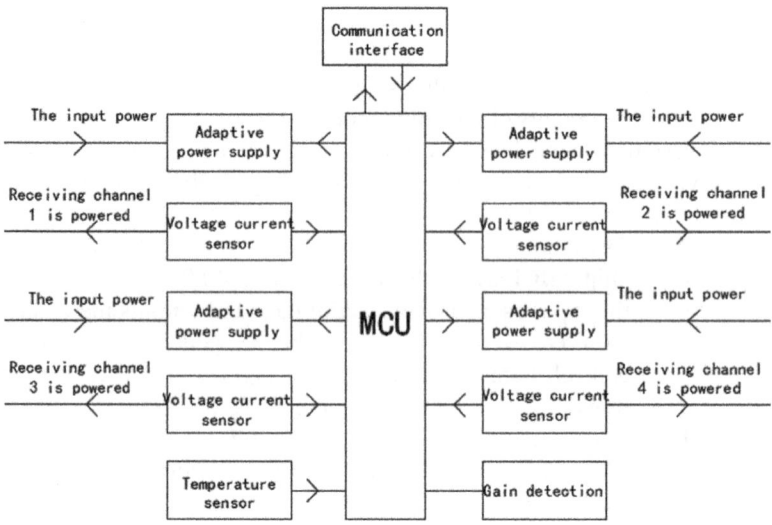

Fig. 4 Adaptive schematic diagram

and performance, power supply topology more flexibility, and by monitoring function can improve the reliability and security of the system, high resolution PWM control function, and can achieve higher power density and faster control loop. This automatic adjustment capability allows for changes in input voltage, output load current, and system temperature, achieving energy savings and superior system performance through dynamic voltage regulation functions and advanced control technology. Thus, the reliability and adaptability of T/R components are further improved.

At the same time, in the T/R module, there are active devices such as LNA, whose gain, output power, standing-wave ratio, and other main RF characteristics will change with the change of environmental temperature. Especially, when working in harsh outdoor environment, its RF characteristics and stability will be seriously affected, which will cause the system to fail to work normally. Generally, the higher the temperature is, the smaller the gain and output power of the components will be. The lower the temperature, the greater the gain and output power of the components. In order to solve this problem, T/R module adopts temperature compensation attenuator for automatic gain compensation, which has the advantages of broadband characteristics, small VSWR, attenuation varying with temperature, and other RF characteristics, so as to effectively compensate the gain, output power fluctuation or RF characteristic drift of active RF devices caused by temperature change [5]. The compensation diagram is shown in Fig. 5 (see Figs. 4 and 5).

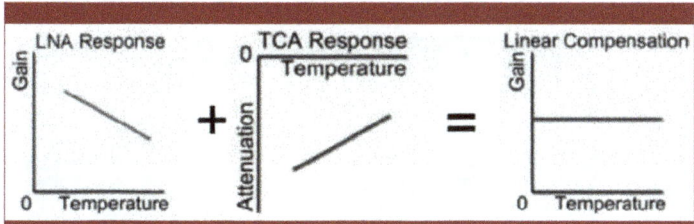

Fig. 5 Gain compensation diagram

3.2 Thermal Design

The external size of the module is 423 mm × 279.5 mm × 41.5 mm. Among them, the heat consumption of the driving amplifier is 5 W, the heat consumption of the final amplifier is 90 W, and the total heat consumption $Q = 95$ W. By calculation, its surface area $S = 2947.6$ cm, the surface heat flux of the module can be calculated as: qfl-f = 95/2947.6 = 0.0322 w/cm^2 (see Fig. 6).

According to the reliability thermal design manual of electronic equipment, under the condition that the heat flux on the system surface is 0.0322 w/cm^2, forced air cooling method with the highest cost performance can be selected. The components conduct forced convection heat dissipation through three fans on the rear panel of the chassis [6].

According to the component, using indoor environment is at an altitude of 4000 m, the environment temperature is −20–50 °C, because the main component heat dissipation issues appeared in the high-temperature working condition, and in calculation, we assume that the environment temperature is 50 °C, the simulation condition is set to the heat conduction, convection and radiation, and the project simulation analysis according to the boundary conditions. In normal working mode, the component temperature distribution cloud after the system reaches equilibrium temperature is shown in Fig. 7.

Fig. 6 Distribution of heat source of components

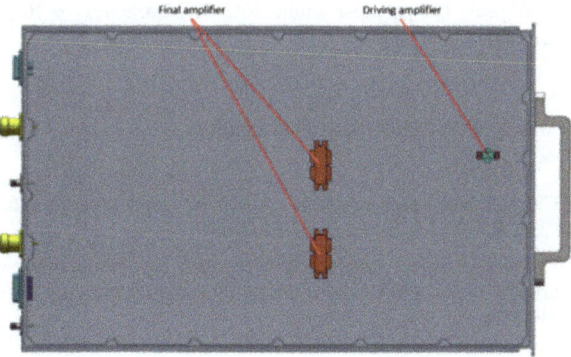

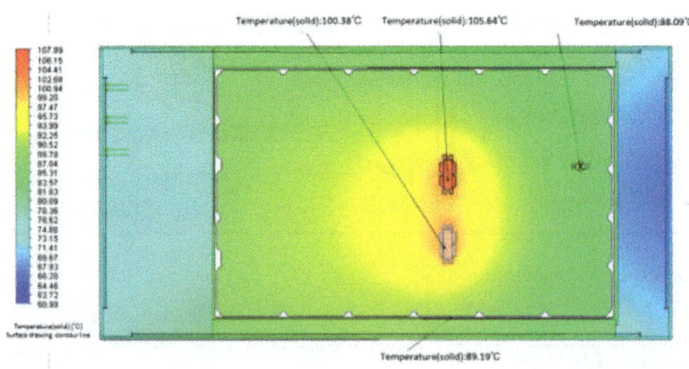

Fig. 7 Component temperature distribution diagram at ambient temperature of 50 °C

According to the data in Tables 1 and 2, the heat resistance of the pushing power amplifier is 1.2 °C/W. Given that the heat consumption of the pushing power amplifier is 5 W, the temperature rise of the pushing power amplifier can be calculated as: 5 W × 1.2 °C/W = 6 °C; the heat resistance of the final power amplifier is 0.15 °C/W. Given that the heat consumption of the final power amplifier is 45 W, the temperature rise of the final power amplifier can be calculated as: 45 W × 0.15 °C/W = 6.75 °C.

The data in Fig. 7 and Tables 1 and 2 can be used to calculate the maximum temperature of a typical device under normal working conditions, cooled by the chassis fan and at an ambient temperature of 50 °C, as given in Table 3.

According to the data in the table, in the worst case, when the ambient temperature is 50 °C, the maximum temperature point of the component is 114.64 °C, according to the temperature of the final power amplifier of the equipment is 114.64 °C, and

Table 1 Thermal resistance of power amplifier

Characteristic	Symbol	Value	Unit
Thermal resistance, junction to case CW: case temperature 80 °C, 25 W CW, 50 V_{dc}, I_{DQ} = 10 mA, 512 MHz	$R_{\theta JC}$	1.2	°C/W
Thermal resistance, junction to case pulse: case temperature 77 °C, 25 W peak, 100 μs pulse width, 20% duty cycle, I_{DQ} = 10 mA, 512 MHz	$Z_{\theta JC}$	0.29	°C/W

Table 2 Heat resistance of final amplifier

Characteristic	Symbol	Value	Unit
Thermal resistance, junction to case CW: case temperature 63 °C, 1250 W CW, I_{DQ} = 100 mA, 230 MHz	$R_{\theta JC}$	0.15	°C/W
Thermal resistance, junction to case pulse: case temperature 66 °C, 1205 W pulse, 100 μs pulse width, 20% duty cycle, I_{DQ} = 100 mA, 230 MHz	$Z_{\theta JC}$	0.03	°C/W

Table 3 Component temperatures at ambient temperature of 50 °C

Temperature monitoring point	Shell temperature °C	Relative ambient temperature rise °C	Junction temperature °C	Location
Outer surface of assembly	89.2	39.2	–	The outer surface is close to the heat source
Final amplifier	107.89	64.64	114.64	The device center
Driving amplifier	88.1	44.1	94.1	The device center

the temperature of the propelling power amplifier is 94.1 °C, all of which are less than the maximum temperature requirements of the device. Therefore, the system can meet the requirements of the environmental temperature of the product. And can work normally under the specified temperature.

4 Physical and Test Results

The external dimension of the component is 423 mm × 279.5 mm × 41.5 mm. The shell material is aluminum–silicon alloy with good thermal conductivity and linear expansion coefficient. T/R module transmit power >63 dBm (≥2000 W).

The minimum output power of T/R module is 63.5 and 63 dBm is 2000 W, so the minimum output power of single channel is more than 2000 W, the minimum output power of double channel is more than 60.5 dBm, and the minimum output power of 60 dBm is 1000 W, so the output power of double channel is more than 1000 W. Figure 8 shows the output power data of the component (see Fig. 8).

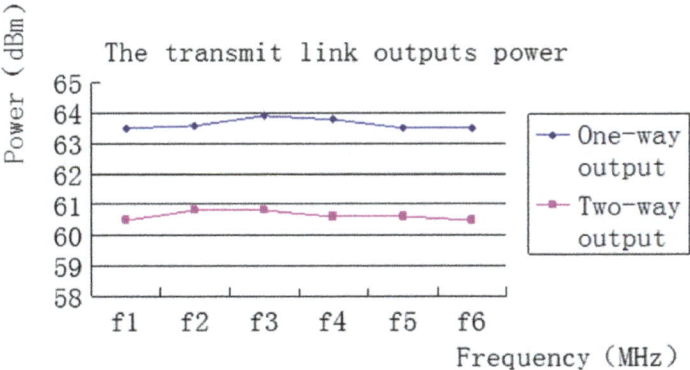

Fig. 8 Output power of transmission link

5 Conclusion

T/R modules designed in this paper for VHF band broadband high-power continuous wave linear power amplifier components, details the design idea, structure, the product of the adaptive technology and thermal design, this paper expounds in detail the two key technologies by means of simulation data comparison and optimization and selects the most appropriate scheme to get the best indicators. This component has finally passed various tests with high reliability and is suitable for communication needs in special areas such as high-altitude areas.

References

1. Rodriguez, J., Lamar, D.G., Miaja, P.F.: Reproducing single-carrier digital modulation schemes for VLC by controlling the first switching harmonic of the DC-DC power converter output voltage ripple. IEEE Trans. Power Electron. **33**(9), 7994–8010 (2018)
2. JEDEC Standard Serial Interface for Data Converters. JEDEC Solid State Technology Association (2012)
3. Jacobson, R.Y., Gupta, T.K., Cheung, P.: Apparatus and method for side printing on low temperature co-fired ceramic (LTCC) substrate: US,US7134391 B2 (2004)
4. Decrossas, E., Glover, M.D., Porter, K.: High-performance and high-data-rate Quasi-Coaxial LTCC vertical interconnect transitions for multichip modules and system-on-package applications. IEEE Trans. Compon., Package Manuf. Technol. **5**(3), 307–313 (2015)
5. Baker, A., Lanagan, M.F., Randall, C.: Integration concepts for the fabrication of TLCC structures. Int. J. Appl. Ceram. Technol. **2**(6), 514–520 (2005)
6. Huang, M.: Research on Millimeter Wave Transceiver Components, pp. 1–5. University of Electronic Science and Technology of China, Chengdu (2008)

Research and Implementation of Point Multiplication Over Elliptic Curve GF(2^m) Based on VLSI

Zhang Qiang and Yang Zhen

Abstract To realize the elliptic curve cryptography (ECC) effectively, the design method of modular multiplication based on optimized binary finite filed algorithm was presented. By the study of the binary finite fields, paralleled modular multiplication algorithm and inversion algorithm which was based on Euclidean algorithm were presented. The two algorithms were optimized during the process and then realized the fast evaluation of point multiplication by adopting Montgomery algorithm. ECC hardware implementation design was proposed based on the algorithm, and converted to logic designs using Verilog RTL, and finally, it worked on the XC7A100T FPGA platform of Xilinx. By pre-simulation, synthetical verification, and analyzing the results of post-simulation, the clock frequency of the designed circuit could reach up to 110 MHz and the operating rate attained to 2.92 ms which demonstrated the feasibility and effectiveness of the project.

1 Introduction

ECC is widely used in hardware-implemented encryption systems, due to its small key length, high security strength, and easy hardware implementation [1]. The point multiplication operation is a prerequisite for achieving the efficiency of ECC, and the effective implementation of finite field arithmetic operations is the key to point multiplication [2]. So far, point multiplication algorithm is mainly implemented by designing the underlying finite field arithmetic unit separately, and then use coprocessors or other control circuits to call underlying functional units [3]. Therefore, researching and implementing each module of the domain operation layer effectively is the key to the FPGA-based ECC point multiplication. Considering the characteristics and advantages of hardware circuit implementation, this article uses ECC of GF(2^m) as the hardware implementation scheme and Montgomery multiplication algorithm. In order to design circuit structure for point multiplication, we design

Z. Qiang · Y. Zhen (✉)
Zaozhuang University, Zaozhuang 277100, Shandong, China

parallel module multiplication algorithms, improve module inverse algorithms, and achieve good results.

2 Elliptic Curve Encryption System

The Weierstrass equation of $GF(2^m)$ is shown in Formula (1) below:

$$y + x \cdot y = x + ax + ba, b \in GF(2^m), b \neq 0 \tag{1}$$

The curve is called Koblitz curve when $b = 1$. (x, y) is the point coordinate on the elliptic curve. This type of curve is the fastest in the implementation of the ECC. The security problem of ECC is based on the elliptic curve discrete logarithm problem (ECDLP), this is why ECC is superior to public key cryptosystems including RSA. The elliptic curve discrete logarithm problem is: Given a finite field F_q and an elliptic curve $E(F_q)$ defined on its field, if the point on the elliptic curve is known P, it is easy to know $Q = KP$. But in turn, when P and Q is known, it is difficult to know K. In this way, Q can be regarded as a public key, K is the private key [4], This is the principle of elliptic curve's point multiplication. In ECC regulations, the polynomial $f(x)$ is a sparse polynomial, use trinomial or pentanominal. In this paper, we choose $f(x) = x^{233} + x^{74} + 1, m = 233$.

3 Circuit Structure Design for Finite Field Operation

3.1 Analysis and Improvement of Modular Multiplication Algorithm

Modular multiplication is the key to the efficient realization of elliptic curve point multiplication, when polynomial are $A(x) = a_{m-1}x^{m-1} + \ldots + a_1 x + a_0$ and
$B(x) = b_{m-1}x^{m-1} + \ldots + b_1 x + b_0$, a, b $\in GF(2^m)$, the product formula of A·B is

$$P(x) = A(x) \times B(x) \bmod F(x) = \sum_{i=0}^{m-1} \sum_{j=0}^{m-1} a_i b_j x^{i+j} \bmod F(x) \tag{2}$$

$F(x)$ is irreducible polynomial in field.

The process of modular multiplication includes two processes: multiplication and modulo. In tradition, multiplication is to multiply two m-bit operands, then modular $f(x)$. The realization of multiplication in the binary domain mainly includes serial structure, parallel structure, and multi-bit parallel structure combining serial and parallel. In order to improve the efficiency of the modular multiplication algorithm, optimized the shift serial algorithm mentioned in paper [5].

Input: $A(x)=a_{m-1}x^{m-1}+...+a_1x+a_0$, $B(x)=b_{m-1}x^{m-1}+...+b_1x+b_0$, non-zero polynomial
        $f(x)$ with highest bit m.
Output: $a(x)\cdot b(x)$ mod $f(x)$.
1. If $a_0=1$, then C=B, else C=0;
2. i from to n-1:
    2.1 b=(b<<1) mod $f(x)$;
    2.2 if $a_i=1$, then c=c mod b;
3. return c.

This algorithm uses a serial shift method, one clock for one shift operation, then do the next XOR operation. So, m-bit operation requires m clocks, takes a long time. According to the characteristics of binary modular reduction operation, in modulo operation, actually judge the highest value of the modulus. If the most bit is 1, then do $f(x)$ XOR modulus; otherwise, keep the original value. So we can consider adopting parallel thinking, first get the value of bi in step 2.1 in parallel, and then complete the operation in step 2.2 in parallel.

The improved algorithm is as follows:

Input: a, b $\in$ GF(2^m), a$\neq$0, b$\neq$0, polynomial $f(x)$.
Output: c=a $\cdot$ b mod $f(x)$.
1. if $a_0=1$, then C=B, else C=0;
2. i from 1 to n-1:
    2.1 b = b << 1
        if $b_{m-1}=1$, then b=b $\oplus$ $f(x)$;
        else b = b;
    2.2 if $a_i=1$, then c=c $\oplus$ b;
3. return c.

According to the shift method to obtain bi in parallel as given in Table 1, considering the highest position b(233), b_i(233) is both 1, for brevity, skip this item here.

Observing Table 1, there are 232 233-bit bi, only related to b, b_i(0), bi(232).

Table 1 Process of obtaining b_i

Data	233bit		232bit	...	76bit	75bit	74bit	...	3bit	2bit	1bit
B	$b(232)$	$b(231)$			$b(75)$	$b(74)$	$b(73)$		$b(2)$	$b(1)$	$b(0)$
b_1	$b(231)$	$b(230)$			$b(74)$	$b_1(74)$	$b(72)$		$b(1)$	$b(0)$	$b_1(0)$
b_2	$b(230)$	$b(229)$			$b_1(74)$	$b_2(74)$	$b(71)$		$b(0)$	$b_1(0)$	$b_2(0)$
b_3	$b(229)$	$b(228)$			$b_2(74)$	$b_3(74)$	$b(70)$		$b_1(0)$	$b_2(0)$	$b_3(0)$
...	...	...			...	...	...		...	...	...
b_{73}	$b(159)$	$b(158)$			$b_{72}(74)$	$b_{73}(74)$	$b(0)$		$b_{71}(0)$	$b_{72}(0)$	$b_{73}(0)$
b_{74}	$b(158)$	$b(157)$			$b_{73}(74)$	$b_{74}(74)$	$b_1(0)$		$b_{72}(0)$	$b_{73}(0)$	$b_{74}(0)$
b_{75}	$b(157)$	$b(156)$			$b_{74}(74)$	$b_{75}(74)$	$b_2(0)$		$b_{73}(0)$	$b_{74}(0)$	$b_{75}(0)$
...	...	...			...	...	...		...	...	...
b_{158}	$b(74)$	$b_1(74)$			$b_{157}(74)$	$b_{158}(74)$	$b_{85}(0)$		$b_{156}(0)$	$b_{157}(0)$	$b_{158}(0)$
b_{159}	$b_1(74)$	$b_2(74)$			$b_{158}(74)$	$b_{159}(74)$	$b_{86}(0)$		$b_{157}(0)$	$b_{158}(0)$	$b_{159}(0)$
...	...	...			...	...	...		...	...	...
b_{231}	$b_{73}(74)$	$b_{74}(74)$			$b_{230}(74)$	$b_{231}(74)$	$b_{158}(0)$		$b_{229}(0)$	$b_{230}(0)$	$b_{231}(0)$
b_{232}	$b_{74}(74)$	$b_{75}(74)$			$b_{231}(74)$	$b_{232}(74)$	$b_{159}(0)$		$b_{230}(0)$	$b_{231}(0)$	$b_{232}(0)$

（1）   For $b_i(0)$:

$b(232)=1$, then $b_1(0)=1$; $b(232)=0$, then $b_1(0)=0$;

$b_1(232)=1$, then $b_2(0)=1$; $b(232)=0$, then $b_2(0)=0$;

the same;

$b_i(232)=1$, then $b_{i+1}(0)=1$; $b_i(232)=0$, then $b_{i+1}(0)=0$;

so, $b_1(0) \sim b_{158}(0)$ are only related to $b(232) \sim b(75)$, can be obtained in parallel at once; $b_{159}(0) \sim b_{232}(0)$ are related to $b_1(74) \sim b_{74}(74)$.

（2）   For $b_i(74)$:

$b(232)=1$, then $b_1(74)=\sim b(73)$; $b(232)=0$,

then $b_1(74)=b(73)$;

$b_1(232)=1$, then $b_2(74)=\sim b_1(73)=\sim b(72)$; $b_1(232)=0$,

then $b_2(74)=b_1(73)=b(72)$;

the same:

when $i <= 74$:

    $b_i(232)=1$, then $b_i(74)=\sim b(74\text{-}i)$; $b_i(232)=0$, then $b_i(74)=b(74\text{-}i)$;

when $i > 74$:

    $b_i(232)=1$, then $b_i(74)=\sim b_{i\text{-}74+1}(0)$; $b_i(232)=0$, then $b_i(74)=b_{i\text{-}74+1}(0)$;

$b_1(74) \sim b_{158}(74)$ are only related to $b(232) \sim b(75)$, we can get it once; $b_{159}(0) \sim b_{232}(0)$ and $b_{159}(74) \sim b_{232}(74)$ are related to $b_1(74) \sim b_{74}(74)$, while $b_1(74) \sim b_{74}(74)$ are only related to $b(232) \sim b(159)$. It can be seen that the data in the b_i form all depend on the input b. Therefore, all b_i values can be obtained in parallel at one time. Then, store all bi values, to facilitate the XOR operation in the following steps. Finally, the result of modular multiplication is obtained through calculation.

3.2 Analysis and Improvement of Modular Inverse Algorithm in Binary Domain

Modular inverse operation is the most complicated and time-consuming operation in finite field arithmetic. We can simply use the nonzero element 'a' to represent the binary polynomial a(x). In the binary field, the inverse element of the nonzero element 'a' is the only element of the field $GF(2^m)$, we can call 'g.' In the field $GF(2^m)$, $ag = 1$, which is $ag \equiv 1(mod\ f)$. This inverse element is represented by the symbol 'a^{-1} mod f,' and can also be directly represented as 'a^{-1}.' Currently widely used are mainly based on the extended Euclid algorithm and the modular inverse algorithm based on Fermat's little theorem. The algorithm based on Fermat's little theorem is to replace the inversion operation with modular multiplication and modular squaring operations. Completing the modular inverse operation requires $log_2(m-1) + w(m-1)-1$ modular multiplication and $m+1$ modular squaring operations. This will not only increase the design complexity of software and hardware, but at the same time, due to repeated calls to modular multiplication and modular squaring operations during the calculation process, the performance will also be greatly reduced. The extended Euclidean algorithm only involves shifting, judgment, and XOR operations in the operation process, and does not have a large number of multiplication operations, so it is easier to implement on hardware and can be designed in parallel. Therefore, according to the currently widely used Euclidean algorithm, further improvements are made to make it possible to perform inversion and modulus operations at the same time during the operation. In this way, the operation of shift judgment is performed at the same time in the operation process, so the input is the inversion operation of m bits, which can be completed in at most 2 m clock cycles, thereby greatly improving the operation efficiency. The modular inverse algorithm used in this paper is described as follows:

Input: non-zero polynomial a(x), b(x) with degree not higher than m and reduced polynomial f(x).

Output: $a^{-1} \bmod f(x)$.

1. u=a, v=f;

2. If mode=0, then x1=1; if mode=1, then x1=b; x2=0.

3. When u≠1 and v≠1, do:

    3.1 When u is an even number, repeat:

        u=u/2;

        if is an even number, then x1=x1/2;

        else, x1=(x1 $\oplus$ f)>>1;

    3.2 When v is an even number, repeat:

        v=v/2;

        if x2 is an even number, then x2=x2/2;

        else, x2=(x2 $\oplus$ f)>>1.

    3.3 When u≥v , then u=(u $\oplus$ v)>>1;

        if the parity of x1 and x2 are the same, then x1=(x1 $\oplus$ x2)>>1;

        if the parity of x1 and x2 are different, then x1=(x1 $\oplus$ x2 $\oplus$ f)>>1.

      otherwise, v=(v $\oplus$ u)>>1;

        if the parity of x1 and x2 are the same, then x2=(x1 $\oplus$ x2)>>1;

        if the parity of x1 and x2 are different, then x2=(x1 $\oplus$ x2 $\oplus$ f)>>1.

4. If u=1, return x1; else, return x2.

In this algorithm, add a mode control bit—'mode,' when 'mode' = 1, change the initial condition to × 1 = b; × 2 = 0, then we can calculate b/a mod q, and it can be verified by Euclidean traditional algorithm. In this way, the modular division or modular inverse operation can be completed by selecting the value of 'mode,' as shown in step 2: when 'mode' = 0, the modular inverse operation is performed, and when 'mode' = 1, the modular division operation is performed. By combining modular division and modular inverse operations into one operation, two operations can be completed by multiplexing a set of circuits in circuit design, and in this way, it can save a lot of operations to save resources and optimize circuit area.

4   The Realization of Elliptic Curve Point Multiplication on FPGA

Montgomery algorithm to realize the design of elliptic curve point multiplication operation is mainly divided into three processes: ① Preprocessing module; ② Loop calculation module; ③ Coordinate conversion module. According to the Montgomery algorithm characteristics, the calculation of the ordinate y is not involved in the module operation, and the affine coordinates p(x,y) and the binary number $(k_{i-1}, ..., k_1, k_0)_2$ are input.

The preprocessing module mainly completes the initialization of the data and provides data for the input of the next module (see Fig. 1). By input, the affine coordinates (x, y) of point P are to generate projection point P1 (X1, Z1) and call a double-point operation to generate P2 (X2, Z2) for use by the loop module. Since P1 and P2 correspond, the initial value of the abscissa of the projection point P1 is (X1 = x, Z1 = 1), the abscissa of the projection point P2 is got by P1's the point double operation.

The main loop calculation is the most time-consuming module, and it is also the most complicated design module of the control part. It mainly includes the shift unit and the point addition and double-point operations of the group operation layer, which is also the most time-consuming operation unit in the point multiplication operation. When each binary bit k_i of the input k is zero, then (X1, Z1) is the input of the double-point module, (X1, Z1) and (X2, Z2) is the input of the dot plus module.

The main function of the coordinate conversion module is to convert the projected coordinates into affine coordinates again after the modular inverse calculation is completed. Since in the first part of the preprocessing module, the affine coordinates of the points have been converted to projection coordinates in advance, so there is no need to perform modular inverse operations when performing point addition and double-point operations. Only need to perform the inverse transformation of the coordinates after the last modular inverse operation is completed.

According to the characteristics of Montgomery algorithm, since there is no data correlation between point addition and doubling point, point addition and doubling point can be scheduled in parallel, and point addition and doubling point operations can be scheduled at the same time for finite field operations. Therefore, this scheduling process can be interpreted as the scheduling of finite field operations by dot product. Dot multiplication needs to call 1 dot addition and 1 double dot operation to complete one iteration operation, and the main loop part needs to complete m − 1 iterations in total. In finite field scheduling, 'M' is the modular multiplication

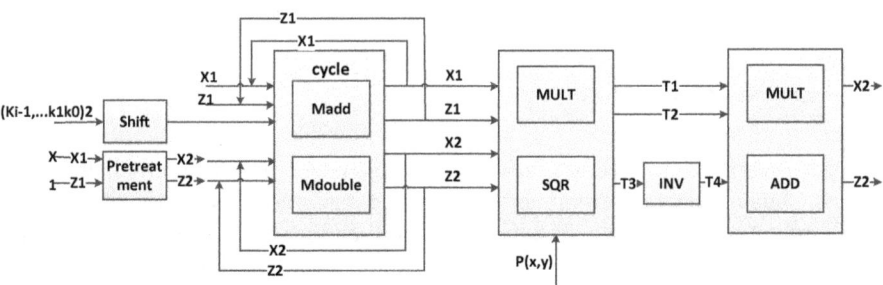

Fig. 1 System structure of elliptic curve point multiplication operation based on Montgomery algorithm

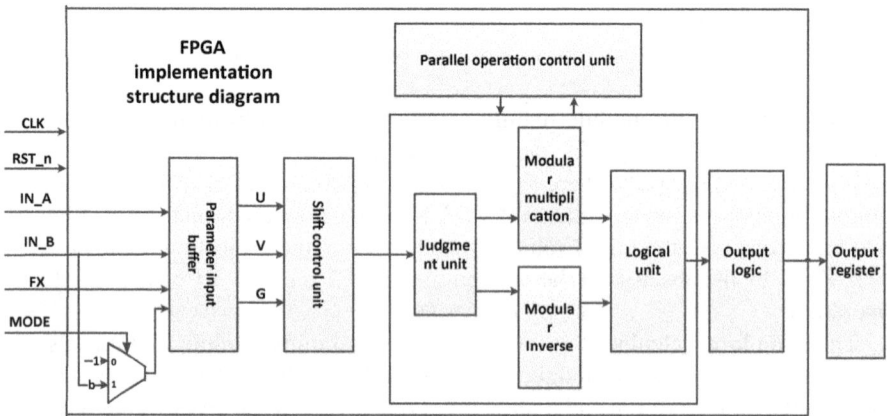

Fig. 2 Structure diagram of finite field operation

time, 'I' is the modular inversion time, and 'AS' is the modular addition time. In the projected coordinates, it takes 6 M + 2AS for a point-add call, and 6 M + 1AS for a double-point call. Therefore, one iteration of the dot multiplication operation calls a total of 12 modular multiplication operations and 3 modular addition operations. After parallel scheduling is used, the completion of one iteration time is equivalent to 6 modular multiplication times (modulo addition time can be ignored). Coupled with a modular inverse operation of the last coordinate inverse transformation, the time to complete a dot multiplication is $6 \text{ M} \cdot (m - 1) + \text{I}$. Taking into account the clock cycles required for modular multiplication and modular inverse realization, a dot multiplication requires $6 \text{ m} \cdot (m - 1) + 2 \text{ m}$ clock cycles. The final output coordinate is the coordinate of scalar multiplication KP (X_2, Z_2).

The circuit structure design of finite field is the key to realize scalar multiplication operation. The improved inversion algorithm has parallelism in the judgment conditions, the calculation steps can be pipelined, and an efficient parallel execution structure is designed based on FPGA. The realized hardware structure is shown below (see Fig. 2).

5 Analysis of Results

In order to verify the correctness and effectiveness of the circuit structure, this article selects Chuanglong's development board as the hardware verification platform for FPGA prototype verification of the point multiplication circuit. The platform chip is Xilinx's ARTIX 7 -XC7A100T (FGG484). Using Verilog for RTL-level model code description, and use Xilinx's Vivado 2015 development software for compilation, synthesis, P&R, and timing analysis, Testbench is written on modelsim10.1, and the test stimulus was input for functional debugging simulation. After synthesis and

P&R, perform corresponding timing constraints. Under the premise of meeting the establishment and holding time of the internal registers of the circuit and no timing violations, the clock frequency can reach up to 110 MHz, and the output results are completely correct. When m = 233, according to the number of clock cycles obtained in the previous article, the time to complete once point multiplication operation can be calculated as 2.92 ms.

At the same time, in order to verify the logic function and timing of the finite field module, the parameter that we used is an elliptic curve with a length of 233 in the GF(2^m) recommended by NIST. Figures 3 and 4 are the simulation results of the modular inverse module and the scalar multiplication module, respectively.

① Modular inverse simulation test:

vector input: a_in; output result : c_out.

② Scalar multiplication module test:

vector input: k, x0_in and y0_in; output results: x_out and y_out. k=6;

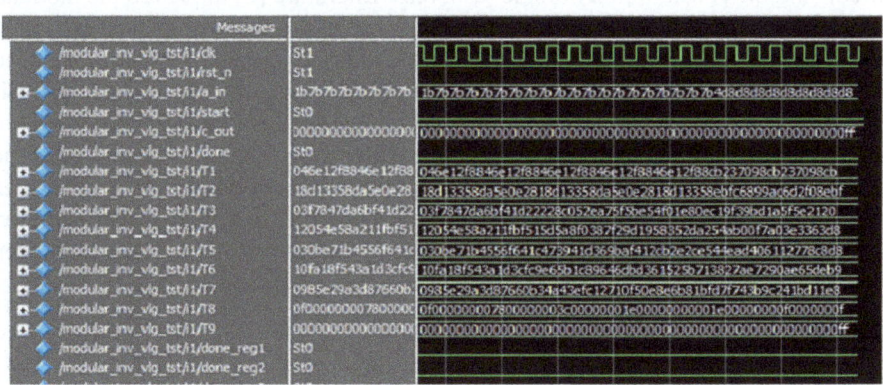

Fig. 3 Modular inversion waveform diagram

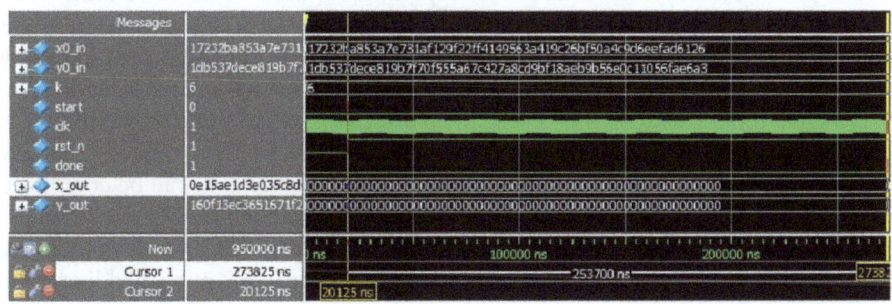

Fig. 4 Scalar multiplication simulation waveform

Table 2 FPGA design speed comparison

Finite field		Paper	FPGA	F(MHz)/T(ms)	SIZE/LUTS
GF(2^{233})		This paper	Xilinx Artix 7	110/2.92	9302
GF(2^{233})	Paper [3]		Cyclone IV EP4CE115	82/4.50	6000
GF(2^{233})	Paper [4]		XCV2000E-7	108/1.62	13,799
GF(2^{233})	Paper [5]		EP2S90F1508C	NA/2.007	15,280

Table 2 lists the experimental data comparison between this design and some other literature studies in recent years, mainly including the time required to complete a point multiplication and the consumption of logic resources. By using Synopsys' design compiler for post-simulation synthesis and checking the comprehensive report of the system design, it can be seen that the design shares 9302 LUTs, approximately 4651 slices. By consulting and comparing related documents, although the hardware platforms implemented by each document are different, we can still compare the design's running time, area scale, and domain to analyze the execution speed of the circuit and the consumption of logic resources. Literature [6] improved the modular multiplication algorithm to optimize the area, but at the expense of the realization efficiency of the dot multiplication. Under the clock frequency of 82 M, the speed of the original design increased by 54%. Literature [7] uses the parallel decomposition scheduling operation of dot multiplication, but does not propose the algorithm optimization of the domain operation level. Although it has an advantage over the speed of the original design, the hardware resource consumption is 48% more than the original design. Compared with other designs, this design has also achieved a good compromise between area and speed, resource consumption has also been significantly improved, and the economics are more prominent.

6 End

In this paper, by studying the arithmetic operation of the elliptic curve finite field, according to the sparse characteristics of ECC polynomials, the parallel modular multiplication idea is proposed, and the traditional Euclidean right shift inversion algorithm is improved, so that it can perform modular inversion and modular division at the same time. Based on this research, an effective dot multiplication circuit was designed and FPGA prototype verification was performed. Experimental results show that when the data bit width is 233 bits, the scalar multiplication can be completed in only 2.92 ms. Compared with other algorithms, this design greatly improves the calculation speed, reduces resource consumption, and achieves a good compromise between area and speed. The cost performance is higher in the published literature. At the same time, as long as we modify the input parameter values of the circuit design, we can achieve the requirements of other different bit-width data, ensuring the validity, versatility, and researchability of the design.

References

1. Li, Y., Yin, X., Shao, M.: Sliding window elliptic curve multiscalar multiplication algorithm based on multi-base representation. Comput. Modernization **281**(01), 15–20 (2019)
2. Zhao, H., Qian, L., Jin, L.: A new McEliece cryptosystem based on subfield subcode of elliptic curve code. Comput. Appl. Softw. **36**(4), 317–322 (2019)
3. Cao, Z.: Scalar Multiplication Design and FPGA Implementation for ECC on GF(2^n). Wuhan University of Technology (2014)
4. Urbano-Molano, F.A., Trujillo-Olaya, V., Velasco-Medina, J.: Design of an elliptic curve cryptoprocessor using optimal normal basis over gf (2 233). In: 2013 IEEE 4th Latin American Symposium on Circuits and Systems (LASCAS), pp. 1–4. IEEE (2013)
5. Tian, Y., Liu, A., Shen, W.C.: An improved scheme of scalar multiplication based on comb method for ellipse curve cryptography. Microelectron. Comput. **32**(5), 99–103 (2015)
6. Liu, H.: Research on Hybrid Encryption Technology Based on Elliptic Curve Cryptosystem and AES. Harbin Engineering University (2019)
7. Wang, Y., Wang, J., Huang, S.: Hardware implementation of a point multiplication algorithm based on parallel scheduling. The J. New Industrialization **4**(7), 20–27 (2014)

Non-contact Hand Sanitizer Auxiliary Device for Public Places

Wenmei Jiang, Ziyue Luo, Meifeng Huang, and Aixuan Huang

Abstract This paper designs and makes a non-contact smart hand sanitizer auxiliary device for public places using the controller Arduino. This device senses the human body's hand through infrared, and the sensing distance is 2 cm. Once hand is detected, the device will automatically turn on the relay and pump out a fixed weight of hand sanitizer. The device can detect the weight of the remaining hand sanitizer through the pressure sensor and upload the collected data to the cloud platform. After receiving the information, the service persons of public places can replenish the hand sanitizer in time as needed. The design has strong real-time performance, stable system, low cost, easy upgrade and improvement and has certain application value.

1 Introduction

With the development of modern civilization and people's demand for environmental protection and health, hand sanitizer is widely used as a necessity in daily life. Especially in public places, the demand for hand sanitizer is greater and more frequent. Using traditional press-type bottle of hand sanitizer or block hand soap will cause bacteria on the hands to remain on the bottle or soap, which may produce cross-infection, especially in the new coronavirus epidemic, when hand washing is an important self-protective work [1]. The traditional contact washing device has caused many people to reject the use of public hand sanitizer, so how to obtain hand sanitizer without contact and send a prompt message for service staffs to remotely judge when the hand sanitizer stock is less than a certain set value. These are the important contents of this paper.

W. Jiang (✉) · Z. Luo · M. Huang · A. Huang
City College of Dongguan University of Technology, Dongguan 523419, China

© The Author(s), under exclusive license to Springer Nature Singapore Pte Ltd. 2021   231
L. C. Jain et al. (eds.), *3D Imaging Technologies—Multidimensional Signal Processing and Deep Learning*, Smart Innovation, Systems and Technologies 236,
https://doi.org/10.1007/978-981-16-3180-1_28

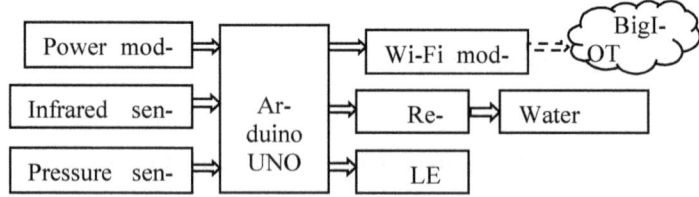

Fig. 1 System box

2 System Design

The system is mainly composed of modules such as Arduino UNO, infrared sensor LM393, 5 kg pressure sensor, Wi-Fi module ESP8266, relay and micro water pump. Among them, Arduino is the central control unit of the entire system, responsible for sensor information collection, data processing and uploading. The function block diagram of the system is shown in Fig. 1.

The working process of the system is as follows: After the hardware circuit is powered on, the initialization of each module is first carried out, including connecting to the wireless network and establishing a communication link with the Big-IOT cloud platform. After the initialization is completed, the infrared sensor module starts to work to determine whether a human body approaches the hand sanitizer device. When the user's hand is close to the infrared sensor module, the module outputs a low-level signal to the digital IO port of the Arduino board. At this time, the micro-pump is controlled by the relay to automatically squeeze a certain amount of hand sanitizer for users to wash their hands, and keep the relay closed when nobody is approaching [2, 3]. Every time after squeezing out the hand sanitizer, the system will use the pressure sensor to automatically detect the weight of the hand sanitizer remaining in the bottle to determine whether it needs to be refilled. If the hand sanitizer is sufficient the green light is on, otherwise the red light is on. At the same time, the weight of the hand sanitizer detected is uploaded to the Big-IOT cloud platform by the Wi-Fi module, and the weight of the remaining hand sanitizer is displayed online in real time. Figure 2 is a flow chart of the hand sanitizer assistance system.

3 Hardware Circuit Design

The complete schematic diagram of the circuit hardware design is shown in Fig. 3, and the system is simulated in the Proteus software. The control module Arduino can perceive the environment through various sensors, and feedback and influence the environment by controlling lights, motors and other devices.

While the system is powered on, the Arduino controller provides the SCK signal required for IIC communication to the infrared sensor HX711 through pin 2 of the

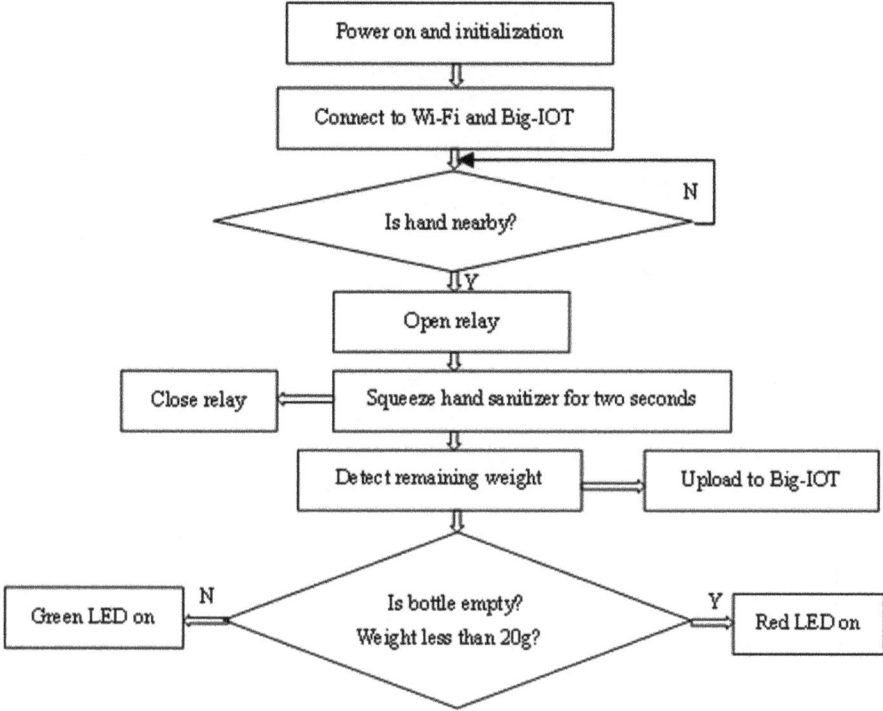

Fig. 2 Main program flow chart

digital IO port, and the infrared sensor sends the detection data to the controller through the analog input port A0 for processing [4]. The controller outputs a low level or a high level to the relay according to the result of infrared sensing. The detection data of the pressure sensor module LM393 is sent to the controller through pin 13 of the digital IO port. The Wi-Fi module ESP8266 and the controller use serial communication for data transmission, and the transmission baud rate is 9600B. The controller's No. 0 and No. 1 digital IO pins are the pre-set serial communication interfaces, which are connected to the RX and TX of the ESP8266 module. The red and green indicator lights on the device shell are also controlled by the digital IO port of the controller, and the LED will be on when the output is high.

3.1 Pressure Sensor Circuit

The pressure sensor consists of two strain gauges and AD unit Hx711. The pressure generated by the weight of the remaining hand sanitizer causes the strain resistance value in the strain gauge to change, thereby outputting the corresponding voltage value. The calculation method of the voltage value is shown in Eq. 1:

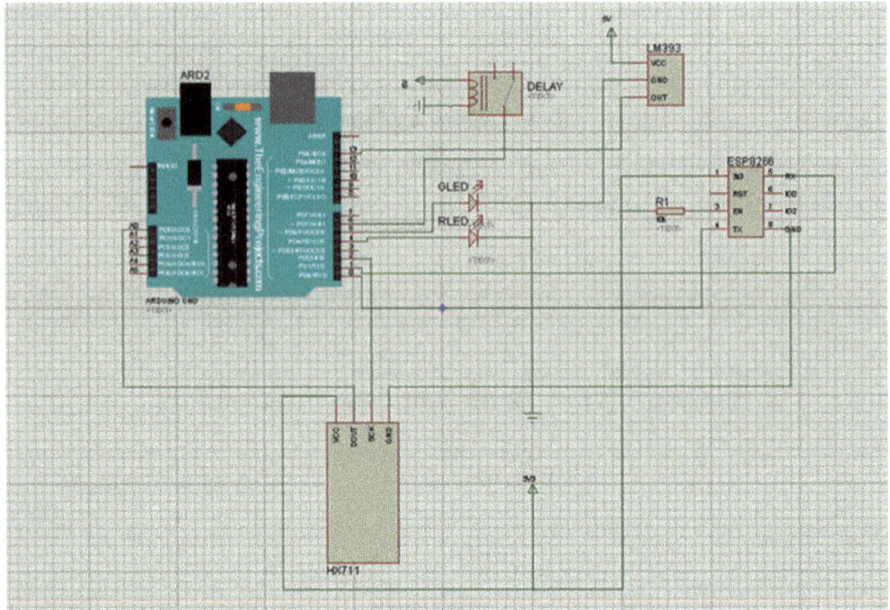

Fig. 3 Circuit hardware design schematic

Full - scale output voltage

$$= \text{working voltage}(5\,\text{V}) * \text{sensitivity}(1.0\,\text{mV/V}) \tag{1}$$

Thus, when there is 5 kg gravity, the output voltage is 5 mV. The AD unit of the pressure sensor selects the A channel working mode, amplifies the output voltage of the strain gauge by 128 times, and then performs AD conversion on the voltage value. The Arduino reads the converted data through the digital IO port.

3.2 Infrared Sensor Circuit

The infrared sensor module LM393 has a pair of infrared emitting tubes and receiving tubes. When the infrared rays emitted by the emitting tube encounter obstacles, they are reflected back and received by the receiving tube, and sent to the comparator circuit for processing, after which the module outputs a low-level signal. The detection distance of the sensor is 2–30 cm, which is adjusted by a potentiometer [5]. The detection angle is 35°.

Table 1 Correspondence between ESP8266 and Arduino

ESP8266	Arduino Uno
UTXD	1 (TX1)
GND	GND
CH_PD/EN	10 kΩ Pull-up resistor to 3.3 V
VCC	3.3 V
UTXD	0 (RX0)

3.3 Wi-Fi Module

Since this system is designed for the hand sanitizer in public restrooms, in order to facilitate the cleaning service personnel to remotely check the remaining amount of hand sanitizer, the design sends the detected weight of the remaining hand sanitizer to the Big-IOT cloud platform through the wireless network. This part of the function is realized by the Wi-Fi module ESP8266. ESP8266 is an ultra-low-power serial Wi-Fi transparent transmission module designed for mobile devices and Internet of Things applications. It can connect users' physical devices to wireless networks for networked communication. The pin connection relationship between ESP8266 and Arduino is shown in Table 1.

4 Results Display and Analysis

We use a 3D printer to make the shell of this non-contact smart hand sanitizer auxiliary device, place the circuit board inside it as shown in Fig. 4, and inlay the hand sanitizer outlet and indicator lights on the front of the shell so that users can take hand sanitizer contactless and the cleaning service personnel can judge the remaining hand washing

Fig. 4 The inside of the hand sanitizer assist device

Fig. 5 Physical image of hand sanitizer assist device

directly. The physical map of the hand sanitizer auxiliary device is shown in Fig. 5. The red indicator light on the panel on indicates that the hand sanitizer is about to be used up, and the green indicator light on indicates that the hand sanitizer is sufficient.

In addition to judging whether hand sanitizer needs to be refilled on the device panel, the device can also observe the weight of the remaining hand sanitizer on the cloud platform. We approached the hand sanitizer several times to pump out the hand sanitizer. From the Big-IOT, we can see that the weight of the hand sanitizer is dropping regularly, which proves that the real-time data upload is successful. At the same time, the weight of the hand sanitizer is output to the serial port monitoring window of the Arduino IDE. Compared with the data in the cloud platform, the results are exactly the same, indicating that the data is uploaded in real time and correctly.

5 Conclusion

This design uses Arduino controller, infrared sensor, pressure sensor and other modules to realize a non-contact hand sanitizer auxiliary device for public places. The device can automatically pump out hand sanitizer without pressing and automatically detect the remaining amount after pumping hand sanitizer. The measured weight of the hand sanitizer is uploaded to the cloud platform in real time, and the cleaning service personnel can check whether it is necessary to replenish the hand sanitizer at any time through the web page or We-Chat applet. The device has high practical value, wide audience and simple structure. It realizes the liquid automatically out and low carbon which is good for environmental protection and effective antibacterial, and it is beneficial to protecting people's health.

References

1. Yang L.H., Zhang Y., Zhao R., Zhang D.Y., Wang P., Hand cleaner application and quality analysis in public place. Chin. J. Health Lab. Technol., **19**, 2859–2860, 2864 (2016).
2. Zheng, C., Deng, J.J., Wang, C.Z., Zhang, X.Z.: Design of automatic dosing device for flotation based on peristaltic pump. Min. Res. Dev. **03**, 128–131 (2019)
3. Li, N.N., Zhang, D.Q., Sun, M., Hu, J.H., Wang, Q., Cui, Z.W.: Intelligent access device for public places. Comput. Knowl. Technol. **03**, 252–254 (2020)
4. Li, S.Q., Zeng, R.X.: Design of data acquisition system of infrared sensor based on STM32. Ind. Control Comput. **08**, 30–31, 33 (2018).
5. Gao, H., Chen, H.C.: The principle and application of infrared technology. Intell. Sens. **05**, 80–82 (2019)

Simulation of Young's Double Slit Interference Experiment Based on MATLAB

Shunrong Chen, Yuchao Jia, Yuanfangzhou Wang, Yonghang Tai, and Yawen Zou

Abstract Due to the high requirements of optical experimental equipment for double slit interference experiment, it is difficult for the laboratory to obtain the theoretical results of the experiment. In this paper, MATLAB software is used to simulate the experimental phenomenon of Young's double slit interference, and the corresponding color interference pattern and light intensity curve can be obtained by inputting different monochromatic light experimental parameters. At the same time, the interference fringes of seven colors of visible light are simulated, and the simulation diagram of the influence of wavelength on the interference fringe spacing is obtained. Finally, the simulation of Young's double slit interference experiment phenomenon is realized, so as to facilitate the teaching and research of optical experiments such as double slit interference.

1 Introduction

In the learning and teaching process of Young's double slit interference experiment, we are generally limited to doing experiments in the laboratory. In the ordinary operation experiment of Young's double slit interference, the change of experimental parameters causes the change of interference fringes is not obvious [1], and optical experiments, such as double slit interference experiment have many factors. Therefore, MATLAB simulation can be applied to the simulation of other optical experiments, and inspire us to explore the inspiration of physical optical experiments [2], so as to have a positive role in the study of optical experiments, and also convenient

S. Chen · Y. Wang (✉) · Y. Tai (✉) · Y. Zou
Yunnan Key Laboratory of Optoelectronic Information Technology, Yunnan Normal University, Kunming, China
e-mail: wyfz@ynnu.edu.cn

Y. Tai
e-mail: taiyonghang@126.com

Y. Jia
Yunnan KIRO-CH Photonics Co. LTD., Kunming, China

© The Author(s), under exclusive license to Springer Nature Singapore Pte Ltd. 2021 239
L. C. Jain et al. (eds.), *3D Imaging Technologies—Multidimensional Signal Processing and Deep Learning*, Smart Innovation, Systems and Technologies 236,
https://doi.org/10.1007/978-981-16-3180-1_29

for teaching in a certain sense, it has important value. MATLAB is a commercial mathematics software [3] produced by MathWorks company of the United States. The predecessors of MATLABA simulation of Yang's double slit interference experiment are also very good. Therefore, in a certain sense, it has a positive significance for the study and teaching of the experiment, and has a certain guidance for the feasibility of MATLAB simulation of double slit interference experiment.

2   Interference Experiment

2.1   Basic Principle

In quantum mechanics, double slit experiment is an experiment to demonstrate the wave and particle properties of micro objects such as photons or electrons. The double slit experiment is a kind of "double path experiment," which is a famous optical experiment. Double slit interference is an important theoretical knowledge in the optical part of physics [4]. It occupies an important position in physics and plays an important role in the study of optics. For the interference optical experiment, Thomas Young carried out a light interference experiment in 1801, which realized the interference of ordinary light source [5], namely the famous Young's double hole interference experiment, and confirmed the fluctuation of light for the first time. The double slit interference experiment has a far-reaching impact on the study of physical light. Later history has proved that this experiment can be ranked among the top five classical experiments in the history of physics. The principle diagram of double slit interference is shown in Fig. 1.

   This design is mainly based on Young's double slit interference principle and its related characteristics, and the powerful simulation design function of MATLAB [6] is used to simulate the double slit interference experiment. The light path schematic diagram of Young's double slit interference experiment is shown in Fig. 2.

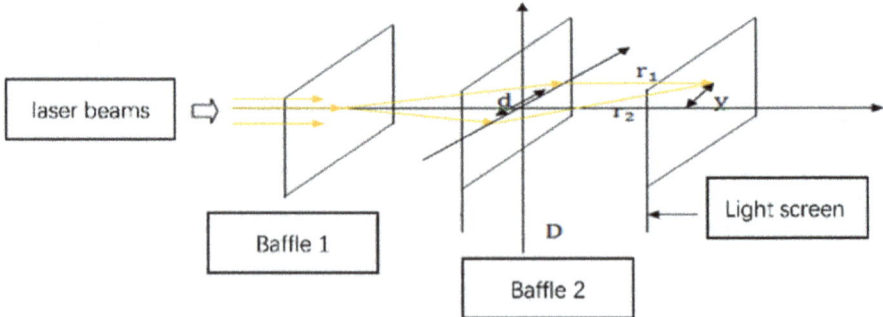

Fig. 1 Schematic diagram of double slit interference experimental device

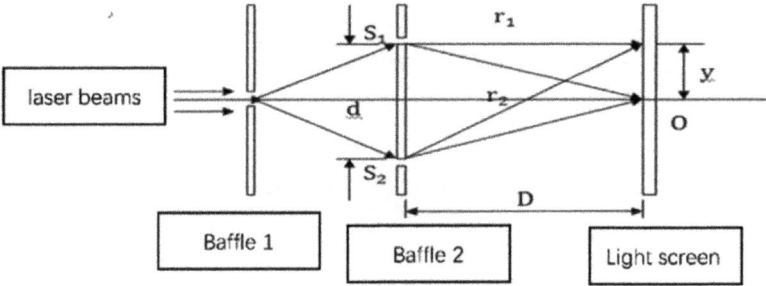

Fig. 2 Light path schematic diagram of Young's double slit interference experiment

2.2 Derivation of Young's Double Slit Interference Formula

According to Fig. 2, light and dark stripes will be generated on the light screen according to the nature of light after passing through two slits. Take two beams of light, for example, the distance between the two slits and the light screen is D, and the distance between the bright stripes on the light screen and the middle distance between the bright stripes on the light screen is y, then $r1, r2$ and δ can be calculated [7], as shown in Formulas (1), (2) and (3):

$$r1 = \sqrt{\left(y - \frac{d}{2}\right)^2 + D^2} \tag{1}$$

$$r2 = \sqrt{\left(y + \frac{d}{2}\right)^2 + D^2} \tag{2}$$

$$\delta = r2 - r1 \approx d \sin\theta = d \cdot \frac{y}{D} \tag{3}$$

Among them, when $r2 - r1 = k\lambda$ is the interference length, when $r2 - r1 = 1/2(2k + 1)$ is the interference cancelation; Though the simulation of software, the position formula of the bright stripe is Formula (4); the formula of the dark stripe position is Formula (5); the distance between two adjacent bright or dark stripes [8] is Formula (6). The formula above will become as follows:

$$y \pm k = \pm k \frac{D}{d} \lambda (k = 1, 2, 3, \ldots) \tag{4}$$

$$y \pm k = \pm (2k - 1) \frac{D}{d} \frac{\lambda}{2} (k = 1, 2, 3, \ldots) \tag{5}$$

$$\Delta y = \frac{D}{d} \lambda \tag{6}$$

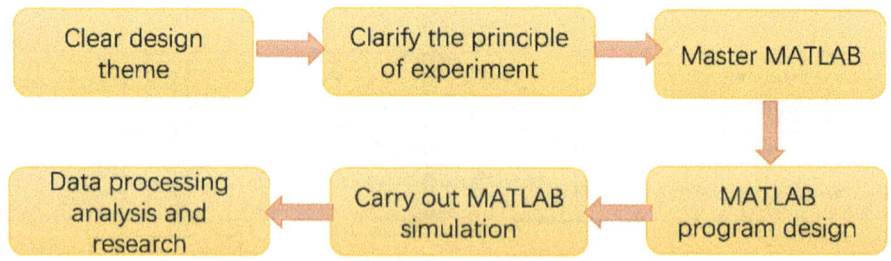

Fig. 3 Simulation design flow of Young's double slit interference experiment

3 Experiment

3.1 Design Scheme of Young's Double Slit Interference Experiment Based on MATLAB Simulation

Young's double slit interference experiment is an important optical experiment in physics. The experiment is mainly completed in the laboratory, and the optical experiment has higher requirements for the external environment [9]. MATLAB is a large-scale scientific computing software produced by MathWorks company of the United States [10]. Through MATLAB language, human–computer interaction can be realized. MATLAB language is used to simulate Young's double slit interference experiment [11]. By inputting different monochromatic light experimental parameters, the corresponding interference pattern and light intensity distribution curve can be obtained, which brings convenience for our study of optical experiment. With the help of MATLAB, the experiment would more perspective and will get more excellent data and information. The simulation design process of Young's double slit interference experiment by MATLAB is shown as Fig. 3.

The specific program design scheme includes: determining the experimental parameters, establishing the theoretical mathematical model, establishing the actual MATLAB language mathematical model, and setting up the interference image scalarization, calculation of light intensity and relative light intensity at any point in the interference image, mapping of monochromatic light wavelength to RGB value of interference image, establishment of chromatographic matrix of interference image, drawing of interference image and drawing of light intensity change curve [12].

3.2 Experimental Results and Analysis

The wavelength λ is 750 nm, the distance d between the double slit is 2 mm, the distance d between the double slit and the light screen is 1 m, the wavelength λ is 585 nm, the distance d between the double slit and the light screen is 2 mm, and the color pattern and light intensity change curve of monochromatic light with the distance from the double slit to the light screen of 1 m are shown in Fig. 4.

The simulation diagram of the distance between the double slit and the light screen is 1 m and 1.2 m, respectively, when the wavelength is 750 nm, the distance between the two seams is 2 mm, and the distance between the two seams and the light screen is 1 m and 1.2 m, respectively. The color interference pattern and intensity change of red light are shown in Figs. 5 and 6.

When the wavelength λ is 750 nm, baffle to double slit distance is 1 m, double slit distance is 2 mm, 3 mm, respectively, Young's double slit interference phenomenon simulation diagram is shown in Fig. 7.

The following is a comparative study on the adjacent fringe spacing of seven monochromatic lights in visible light when the distance between two slits is 2 mm and the distance between double slit and light screen is 1 m. The detail is shown in Table 1.

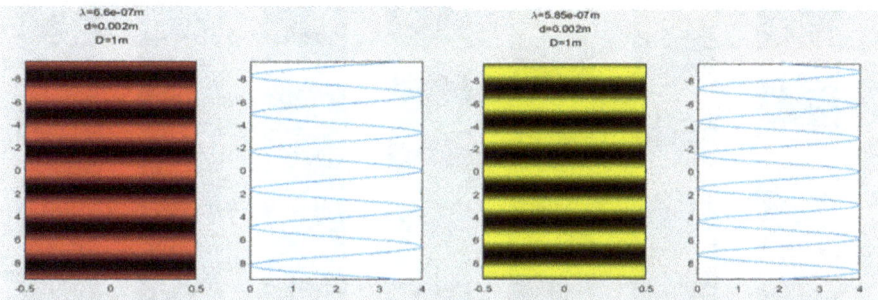

Fig. 4 Color interference pattern and light intensity change diagram of monochromatic light

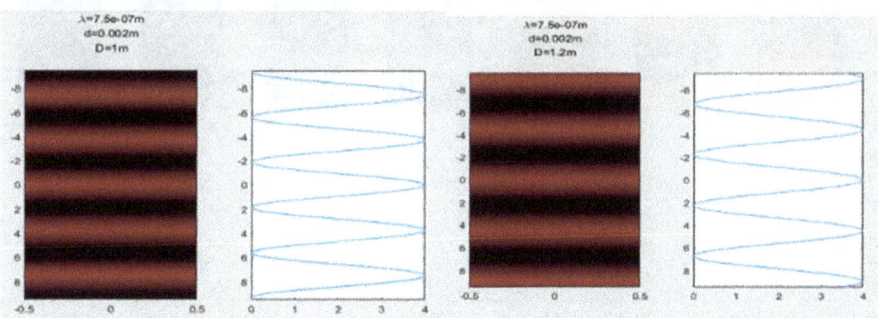

Fig. 5 Color interference pattern and intensity variation of red light with different D

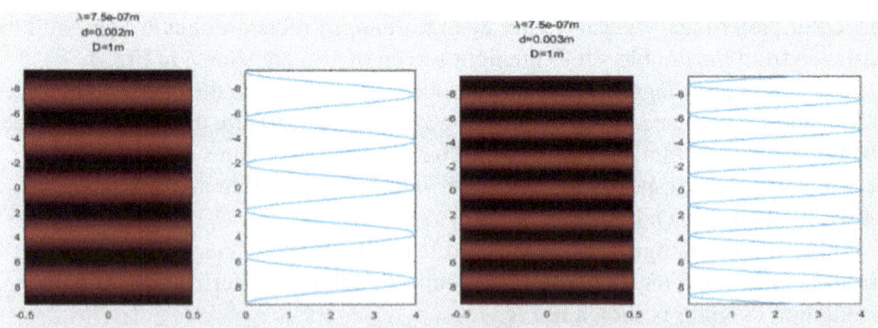

Fig. 6 Color interference pattern and intensity variation of red light with different d

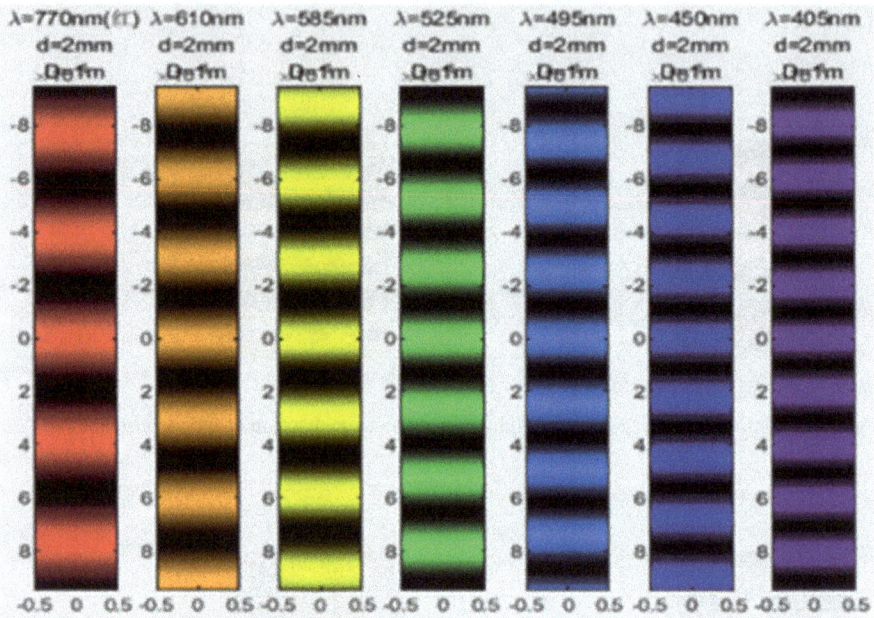

Fig. 7 Simulation of the influence of wavelength on interference fringe spacing

Table 1 Visible interference fringe spacing

Visible light	Red	Orange	Yellow	Green	Cyan	Blue	Purple
Wavelength (nm)	760	610	585	525	495	450	405
Fringe spacing (um)	380	305	292.5	262.5	247.5	225	202.5

4 Conclusions

In this experiment, Young's double slit interference experiment is introduced. Combined with the function of MATLAB software, the phenomenon of Young's double slit interference experiment is simulated by using MATLAB. MATLAB programming realizes that the corresponding color interference pattern, interference image chromatographic matrix and light intensity change curve can be obtained by inputting different monochromatic light experimental parameters. Through the simulation pattern, the influence of wavelength on the interference fringe spacing can be clearly seen, which has certain value for the research of interference pattern. Therefore, through the simulation of Young's double slit interference experiment results, the double slit interference experiment is clearer and more intuitive, which makes the students easier to understand and has a positive impact on the research of physical optics to a certain extent.

References

1. Frabboni, S., Gazzadi, G.C., Pozzi, G.: Young's double-slit interference experiment with electrons. Am. J. Phys. **75**(11), 1053–1055 (2007)
2. Xiangechnology, C.: Regular research of light interference in SPR phenomenon. Instrum. Tech. Sens. **6**, 41–42+52 (2004)
3. Chen, Z., Pu, J.: Degree of polarization in Young's double-slit interference experiment formed by stochastic electromagnetic beams. JOSA A **24**(7), 2043–2048 (2007)
4. Pu, J., Cai, C., Nemoto, S.: Spectral anomalies in Young's double-slit interference experiment. Opt. Express **12**(21), 5131–5139 (2004)
5. Lianzhou, C.Z.Z.G.R., Jixiong, P.: Determining the orbital angular momentum of vortex beam by Young's double-slit interference experiment. Chin. J. Lasers **7**, 1063–1067 (2008)
6. Grant, M., Boyd, S., Ye, Y.: CVX: Matlab Software for Disciplined Convex Programming (2008)
7. Shu, J.H., Chen, Z.Y., Liu, Y.X., Pu, J.X.: Spectral anomalies of Young's double-slit interference experiment with vortex beam. Opt. Tech. **35**, 1–5 (2009)
8. Anand, S., Kamalasanan, M.N.: Experimental study of the spectral anomalies in Young's interference experiment. Opt. Commun. **266**(2), 469–472 (2006)
9. Yu-ming, M.A.O.: Computer simulation of young's double slit experiments based on MATLAB. J. Hubei Normal Univ. (Nat. Sci.) **1**, 21–24 (2007)
10. Cheng, L., Xu, D.D.: Application of Matlab simulation in communication principle teaching. Res. Explor. Lab. **2**, 122–124 (2010)
11. Etter, D.M., Kuncicky, D.C., Hull, D.W.: Introduction to MATLAB. Prentice Hall (2002)
12. Seiler, M.C., Seiler, F.A.: Numerical recipes in C: the art of scientific computing. Risk Anal. **9**(3), 415–416 (1989)

Simulation of Impact of $CH_3NH_3PbI_3$ Based Thin Film Transistor Polarization

Xianrong Mao, Qiuhong Tan, Chao Zhang, Li Ren, Xia Zhang, and Jiyu Zhao

Abstract The ferroelectric memory using ferroelectric field effect transistors as the basic unit has received more and more attention because of the increasing demands on memory due to the development of electronic products. Non-volatile memory based on ferroelectric field effect transistors is considered to be one of the most promising next-generation memories. This subject selects lead methyl iodide ($CH_3NH_3PbI_3$)-based ferroelectric field effect transistor as the research object, and uses MATLAB software to simulate the effect of the polarization strength of the $CH_3NH_3PbI_3$-based ferroelectric field effect transistor on the memory window on the basis of predecessors. The simulation results show that the larger the storage window, and the faster the maximum value of the storage window, the better the storage characteristics and the next-generation 3D imaging technology.

1 Introduction

Due to the advantages of non-destructive read–write operation, low-power consumption and high integration, ferroelectric gate field effect transistors used in future nonvolatile random-access memories have attracted wide attention. Various studies began in 1960 AD, but basically all the studies used silicon as semiconductor. Therefore, it is proposed to use silicon semiconductor materials instead of oxide-based semiconductor materials, and because of the good interface effect between ferroelectric layer and semiconductor layer, researchers try to use this TFT (Thin Film Transistor) structure to manufacture ferroelectric memory, so as to obtain better retention performance. Oxide semiconductor channel transistors usually operate in a storage depletion mode because it is difficult to control the conductivity type by obtaining channel inversion through ion implantation in oxide semiconductors [1].

X. Mao · Q. Tan (✉) · C. Zhang (✉) · L. Ren · X. Zhang · J. Zhao
School of Physics and Electronic Information, Yunnan Normal University, Kunming, China
e-mail: tanqiuhong1@126.com

C. Zhang
e-mail: zhchynnu@foxmail.com

© The Author(s), under exclusive license to Springer Nature Singapore Pte Ltd. 2021   247
L. C. Jain et al. (eds.), *3D Imaging Technologies—Multidimensional Signal Processing and Deep Learning*, Smart Innovation, Systems and Technologies 236,
https://doi.org/10.1007/978-981-16-3180-1_30

Therefore, when the carriers on the oxide semiconductor are exhausted, the Pf on the oxide semiconductor becomes unstable [2]. Therefore, the entire dielectric crystal has a polarity called a special polarity direction in a specific direction. Ferroelectric materials are mainly thermoelectric, piezoelectric and ferroelectric [3].

2 Related Work

Ferroelectric memory is a new type of nonvolatile memory. It started earlier and realized industrialization at the earliest. Low-power consumption, fast reading and writing speed and strong radiation resistance. However, ferroelectric memory also has some disadvantages, such as difficulty in improving integration, serious process pollution and incompatibility with conventional complementary metal oxide semiconductor (CMOS) process [4]. It needs further research and solution.

At present, most LCDs use amorphous silicon thin film transistors as the most basic pixel units, such as a-Si and H-TFT. However, they have low-carrier mobility and serious photo-sensitive degradation. Therefore, it is very important to find the best thin film transistor (TFT). As a basic memory cell, ferroelectric field effect transistors have been widely concerned. The insulating layer in conventional MOSFET is replaced by ferroelectric material in ferroelectric field effect transistor. Reading and storing information are realized by controlling the state and position of source-drain channel through polarization reversal in ferroelectric thin film. The principle of reading information by ferroelectric field effect transistor is: a read pulse is added to the source and drain electrodes. Because the ferroelectric thin film has been written with data before, the channel of the source and drain electrodes has been adjusted and controlled. At this time, the stored information can be read only according to the source and drain current [5].

3 Simulation

3.1 Polarization Theory

The electrical properties of $CH_3NH_3PbI_3$-based ferroelectric field effect transistor were calculated by using the combination of the semiconductor transport theory, and the improved model, as elaborated as follows.

Parameter	Description	Parameter value
P_s	Saturation polarization	45 $\mu C/cm^2$
P_r	Remnant polarization	32 $\mu C/cm^2$
E_C	Coercive field	80 kV/cm

(continued)

(continued)

Parameter	Description	Parameter value
ε_f	Ferroelectric dielectric constant	300
t_F	Ferroelectric film thickness	300 nm
μ_n	Substrate electron mobility	10 cm^2/s V
N_D	Substrate doping concentration	10^{16} cm^{-3}
L	Channel length	10 μm
W	Channel width	100 μm
P_Z	ZnO largest spontaneous polarization	5 μC/cm^2

Miller hyperbolic function model: a mathematical model of hysteresis loop suitable for saturated polarization. The mathematical expression is as follows:

$$P_{sat}^+ = P_s \tan h\left(\frac{E - E_c}{2\delta}\right) + \varepsilon_r \varepsilon_0 E \tag{1}$$

$$P_{sat}^- = P_s \tan h\left(\frac{E + E_c}{2\delta}\right) + \varepsilon_r \varepsilon_0 E \tag{2}$$

And $\delta = E_c \left(\ln\left(\frac{1+\frac{P_r}{P_s}}{1-\frac{P_r}{P_s}}\right)\right)^{-1}$, the P_s is saturation polarization, E_c is coercive field, ε is dielectric constant. Hang-Ting Lue model: a mathematical model of hysteresis loop suitable for unsaturated polarization [5]. The mathematical expression is as follows:

$$P^+(E, E_m) = P_s \tan h\left(\frac{E - E_c}{2\delta}\right) + \varepsilon_r \varepsilon_0 E$$
$$+ \frac{1}{2}\left(P_s \tan h\left(\frac{E_m + E_c}{2\delta}\right) - P_s \tan h\left(\frac{E_m - E_c}{2\delta}\right)\right) \tag{3}$$

$$P^-(E, E_m) = P_s \tan h\left(\frac{E + E_c}{2\delta}\right) + \varepsilon_r \varepsilon_0 E$$
$$- \frac{1}{2}\left(P_s \tan h\left(\frac{E_m + E_c}{2\delta}\right) - P_s \tan h\left(\frac{E_m - E_c}{2\delta}\right)\right) \tag{4}$$

$$P_d(E_m) = \varepsilon_r \varepsilon_0 E_m$$
$$+ \frac{1}{2}\left(P_s \tan h\left(\frac{E_m + E_c}{2\delta}\right) + P_s \tan h\left(\frac{E_m - E_c}{2\delta}\right)\right) \tag{5}$$

MATLAB simulation is used to consider the polarization effect of CH$_3$NH$_3$PbI$_3$, and the polarization theoretical results of CH$_3$NH$_3$PbI$_3$-FET can be obtained by using formulas (3), (4) and (5). In this progress, because of the LD, at the same time capacitance of MFMIS transistor:

$$C_F = \frac{\varepsilon_0 \varepsilon_r A_F}{t_F} \tag{6}$$

$$C_I = \frac{\varepsilon_0 \varepsilon_I A_I}{t_I} \tag{7}$$

where C_F is the capacitance of the ferroelectric layer, and C_D is the capacitance of the semiconductor depletion layer,

$$C_D = \frac{A_I \varepsilon_{Si} \varepsilon_0}{\sqrt{2} L_D} \frac{\frac{n_i^2}{N_D^2}(-\exp(-\beta\psi_S)+1)+(\exp(\beta\psi_S)-1)}{\sqrt{\frac{n_i^2}{N_D^2}(\exp(-\beta\psi_S)+\beta\psi_S-1)+(\exp(\beta\psi_S)-\beta\psi_S-1)}} \tag{8}$$

$$C_{total} = \left(\frac{1}{C_F} + \frac{1}{C_I} + \frac{1}{C_D}\right)^{-1} \tag{9}$$

3.2 MATLAB Simulation

According to the C–V characteristics and polarization theory of $CH_3NH_3PbI_3$-FET, the following data are obtained by simulation with MATLAB: keep P_s and P_r unchanged, and the storage window characteristic curve when P_z is changed as shown in Fig. 1.

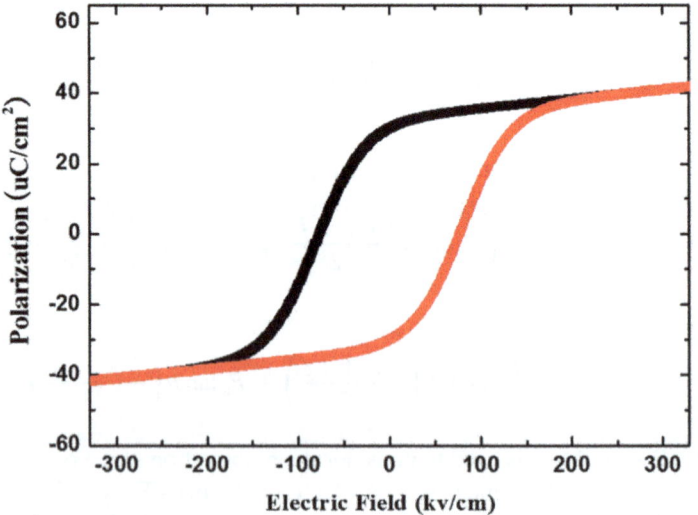

Fig. 1 The P–E hysteresis loop of $CH_3NH_3PbI_3$-FET

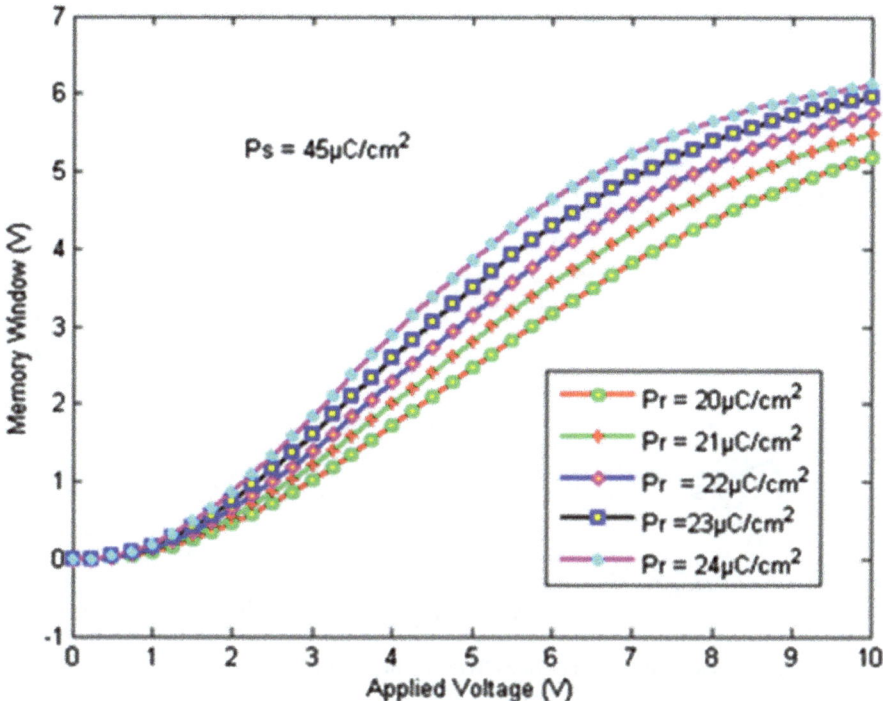

Fig. 2 Memory window dependence of a CH$_3$NH$_3$PbI$_3$-FET on the P_r

Figure 2 is the influence of dielectric constant (abscissa) and saturation polarization intensity of PZT on storage window (ordinate) depicted by MATLAB, in which the applied voltage is kept at v and $P_r/P_s = 19/21$ (Fig. 3).

It can be seen from the capacitance formula that with the increase in E_f, the ferroelectric capacity (ferroelectric capacity) increases, thus the partial voltage at the ferroelectric layer decreases relatively and the storage window decreases.

4　Conclusion

Under typical growth mode, CH$_3$NH$_3$PbI$_3$ thin films usually have spontaneous polarization (P_S) with upward direction and no voltage reversal. Therefore, in the off state, the two polarization directions are opposite, and the polarization of the ferroelectric thin film exhausts the carriers on the surface of methylamine lead iodide. In the state that carriers are contained on the surface of methylamine lead iodide, the polarization of ferroelectric thin films will accumulate, because the two polarization directions are the same, so the polarization of ferroelectric thin films can be enhanced by the

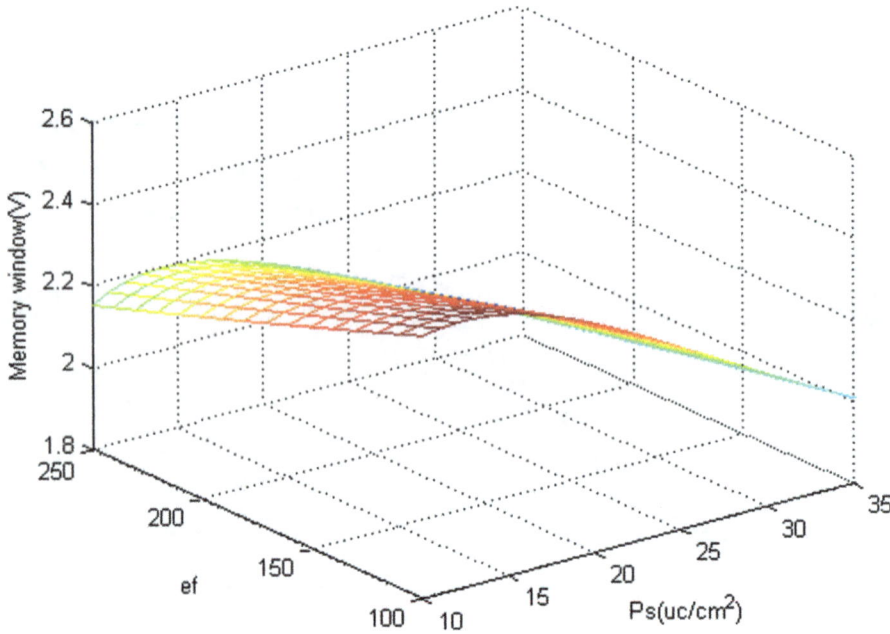

Fig. 3 Memory window dependence of a $CH_3NH_3PbI_3$-FET on the P_s and E_f

inherent polarization of $CH_3NH_3PbI_3$, which can increase the on-state current of $CH_3NH_3PbI_3$-TFT.

The closer the difference between the positive and negative coercive voltages between the memory window and ferroelectric thin film is, the smaller the influence of charge injection effect will be, and the more ideal the barrier layer thickness and defect effect will be [6]. The larger the residual polarization intensity (P_r), the larger the storage window, and the faster it reaches the maximum value of the storage window, which means the better the storage characteristics. This is because, when the residual polarization intensity is high, a higher voltage is needed to reduce it to zero or even reverse it, at which time the storage window can be made larger.

References

1. Ng, K., Hillenius, S.J., Gruverman, A.: Transient nature of negative capacitance in ferroelectric field-effect transistors. Solid State Commun. **265**, 12–14 (2017)
2. Wen, Y., Liu, Y., Guo, Y., Yu, G., Hu, W.: Experimental techniques for the fabrication and characterization of organic thin films for field-effect transistors. Chem. Rev. **111**(5), 3358–3406 (2011)
3. Sata, N., Shibata, Y., Iguchi, F., Yugami, H.: Crystallization process of perovskite type oxide thin films deposited by PLD without substrate heating: Influence of sputtering rate and densification-driven high tensile strain. Solid State Ionics **275**, 14–18 (2015)

4. Sun, D.G., Liu, Z., Huang, Y., Ho, S.T., Towner, D.J., Wessels, B.W.: Performance simulation for ferroelectric thin-film based waveguide electro-optic modulators. Opt. Commun. **255**(4–6), 319–330 (2005)
5. Belous, A.G., V'Yunov, O., Kobylianskaya, S.D., Ishchenko, O.O., Kulinich, A.V.: Thin films of organic-inorganic perovskites $CH_3NH_3PbI_3$: control of microstructure and properties. In: XVI International Conference on Physics and Technology of Thin Films and Nanosystems (2017)
6. V'yunov, O., Belous, A., Kobylianska, S., Kovalenko, L.: Impedance analysis of thin films of organic-inorganic perovskites $CH_3NH_3PbI_3$ with control of microstructure. Nanoscale Res. Lett. **13**(1), 98 (2018)

Preparation and Output Characteristics of Molybdenum Disulfide Field Effect Transistor

Chaorun Pu, Qiuhong Tan, Chao Zhang, Li Ren, Xia Zhang, and Jiyu Zhao

Abstract The two-dimensional material molybdenum disulfide is an n-type semiconductor, so it can be directly used as a semiconductor in field effect transistors. Molybdenum disulfide has a direct band gap of 1.3–1.8 eV for high mobility, low power, and the high switching ratio field effect transistor has a great application space. Molybdenum disulfide has become the research focus of channel materials in silicon-based devices. In this paper, a few layers of molybdenum disulfide were prepared by chemical vapor deposition (CVD) and prepared into molybdenum disulfide field effect transistors by masking technique. The output characteristics were studied and tested to demonstrate the superior performance and application value of molybdenum disulfide in next-generation 3D imaging technology.

1 Introduction

With the development of new semiconductors, two-dimensional materials have become the research focus to replace the channel materials of silicon-based devices. Graphene has the advantages of high mobility, low power consumption, and high-speed operation, but its intrinsic band gap is zero, which makes the current switching ratio of graphene-based field effect transistors extremely low. The result shows the practicability of MoS_2 in switching devices, so molybdenum disulfide has great application value in micro-nanoscale and the preparation of logic circuits and transistors [1]. In the research done by Narae Kang et al., it can be observed that the optical characteristic curve of MoS_2 lamellae treated by oxygen plasma changes [2]. It also reflects that after oxygen plasma treatment, MoO_3 generated makes MoS_2 lamellae become disordered regions. Islam et al. [3] also reported that after oxygen plasma treatment of single-layer MoS_2, MoS_2 will change from semiconductor to insulator,

C. Pu · Q. Tan (✉) · C. Zhang (✉) · L. Ren · X. Zhang · J. Zhao
School of Physics and Electronic Information, Yunnan Normal University, Kunming, China
e-mail: tanqiuhong1@126.com

C. Zhang
e-mail: zhchynnu@foxmail.com

© The Author(s), under exclusive license to Springer Nature Singapore Pte Ltd. 2021   255
L. C. Jain et al. (eds.), *3D Imaging Technologies—Multidimensional Signal Processing and Deep Learning*, Smart Innovation, Systems and Technologies 236,
https://doi.org/10.1007/978-981-16-3180-1_31

and the mobility, on-state current, and resistance of the device will change. At the same time, the photoelectric response capability of the device can be improved by oxygen plasma treatment of multilayered MoS_2 sheets [4].

For the mobility switching ratio of MoS_2 field effect transistor, we can change and increase the mobility by changing the electrode and insulating layer. The mobility and switching ratio are calculated by the transfer characteristic curve. Secondly, improving the performance of MoS_2 field effect transistor also depends on the size of MoS_2. This time, we will use chemical vapor deposition method to obtain the larger molybdenum disulfide with less layers.

2 Preparation of Molybdenum Disulfide

In this experiment, chemical vapor deposition (CVD) will be used to prepare molybdenum disulfide with large size. The instrument used is SDGL-1700-III vacuum tube-type high-temperature furnace in three temperature zones of Shanghai Ju Precision Instrument Manufacturing Co., Ltd. The specific operation steps of the preparation process are as follows:

Preparation: (1) substrate treatment: use silicon wafer of Si/SiO_2, the brighter side of which is SiO_2, cut it into 2 cm in length, 7 mm in width and 300 nm in thickness, and clean the substrate surface with acetone, ethanol, and deionized water in turn for later use; (2) selection of raw materials: prepare two small ceramic boats with a size of 77 mm and clean them with ethanol. One ceramic boat is filled with enough sulfur powder and pressed as full as possible, and the other ceramic boat is evenly distributed with 0.01 g MoO_3; (3) place the substrate (silicon wafer) processed in the first step in the middle of the ceramic boat filled with 0.01 g MoO_3, put the bright side of the silicon wafer down, then put the two ceramic boats into the quartz tube, and put the quartz tube on the vacuum tube high temperature furnace in the three-temperature zone to adjust the distance between the two ceramic boats, so that they are in the middle of the instrument and the distance between them is 18 cm. After calibrating the distance, put the quartz tube into the high-temperature zone of the instrument; (4) preparation of the instrument: before starting the test, check that the valve is closed tightly, turn on the mechanical pump, and slowly pump air, but it is not easy to pump the sulfur powder too fast. After the pointer indicates 0, introduce AR (H/5%) at a gas flow rate of 100 sccm; (5) start setting up the instrument: raise the temperature to 300 °C at a rate of 30 °C/min, stay for 20 min to remove distilled water, raise the temperature to 750 °C at a rate of 37.5 °C/min, keep it for 15 min, then reach 700 °C in 20 min, and then naturally cool it from 700 °C to room temperature. In this process, Ar (h/5%) was introduced at a gas flow rate of 50 sccm.

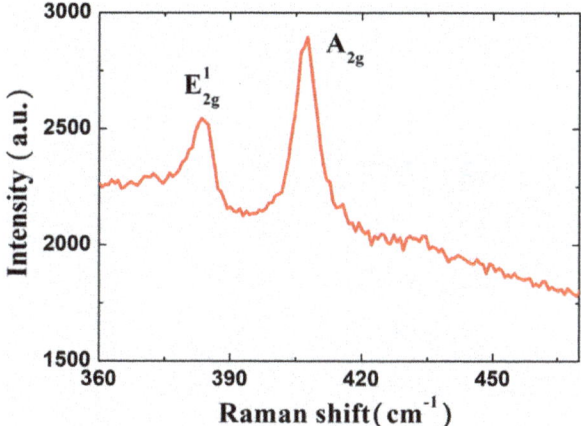

Fig. 1 Raman spectroscopy of molybdenum disulfide (Raman)

2.1 Raman Spectroscopic Characterization

We characterize the prepared molybdenum disulfide by confocal Raman spectroscopy, and the characterization of molybdenum disulfide only needs to be used for the number of layers, thickness, and surface quality of samples. Figure 1 is a Raman shift, and the prepared molybdenum disulfide is characterized on this optical platform.

Raman spectra of molybdenum disulfide are divided into E^1_{2g} and A_{2g}, which represent the basic vibration modes of two different vibration modes [5]. The E^1_{2g} mode is related to the vibration of atoms in the inner layer direction, and the A_{2g} mode is related to the interlayer vibration of atoms in the direction perpendicular to the interface. These two peaks are related to the number of molybdenum disulfide layers.

2.2 Preparation of Molybdenum Disulfide Field Effect Transistors

The molybdenum disulfide prepared by vapor deposition method is used in the semiconductor part of field effect transistor. In this process, the molybdenum disulfide field effect transistor is prepared directly on the original silicon wafer substrate with a thickness of 300 nm instead of transferring the prepared molybdenum disulfide. Field effect transistor is the most basic electronic component in digital logic circuit, which includes semiconductor, source, drain, and gate [6]. We need to plate Au source and drain electrodes on molybdenum disulfide with few layers of silicon substrate, which can directly complete the preparation of molybdenum disulfide field effect transistor.

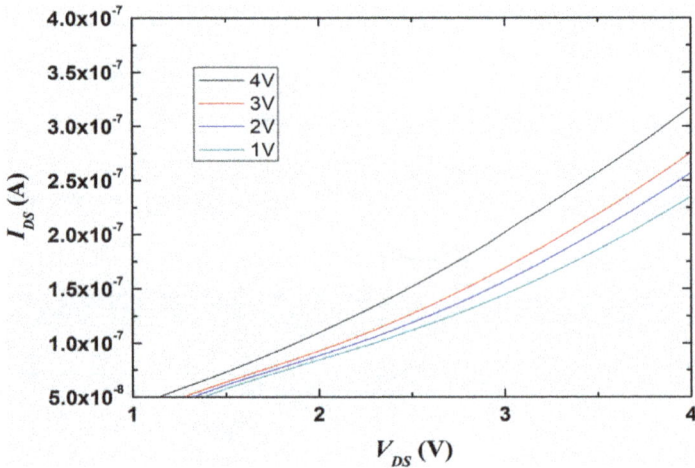

Fig. 2 Output characteristic curve of transistor

3   Optical Microscope Characterization

3.1   Output Characteristics of Transistors

Output characteristic curves of source-drain voltage, gate voltage, and corresponding source-drain current of molybdenum disulfide field effect transistor are shown in Fig. 2. MoS$_2$ field effect transistor shows the typical p-channel transistor characteristics and saturated source-drain currents at low source-drain voltage. When the gate voltage $V_{GS} = -4$ V, and $V_{DS} = 4$ V, the "on" current of MoS$_2$ field effect transistor reaches to 3.2×10^{-7} A. This is because the high-quality MoS$_2$ few layers, which was treated with surface smooth, that lack defects around the MoS$_2$ surface and have improve interface between the MoS$_2$ and SiO$_2$.

4   Conclusion

In this experiment, molybdenum disulfide was prepared by CVD, and few layers of molybdenum disulfide were obtained, which were characterized by optical microscope and Raman spectrum, and the layers and quality of the prepared molybdenum disulfide were verified. It was found that it not only has a good linear relationship and ohmic contact, but also has a high switching ratio, while reducing the power consumption of the device. Molybdenum disulfide, a new semiconductor material studied in this study, has a direct band gap, which shows its superior performance in fabricating field effect transistors. It is believed that more new materials will emerge in the future, and digital logic circuits and 3D imaging technology will become better and better.

References

1. Chee, S.S., Seo, D., Kim, H., Jang, H., Lee, S., Moon, S.P., Lee, K.H., Kim, S.W., Choi, H., Ham, M.H.: Lowering the Schottky barrier height by graphene/Ag electrodes for high-mobility MoS_2 field-effect transistors. Adv. Mater. **31**(2), 1804422.1–1804422.7 (2019)
2. Jahangir, I., Wilson, A., Singh, A.K., Sbrockey, N., Koley, G.: Graphene/InN and Graphene/MoS_2 heterojunctions: characterization and sensing applications. In: 2014 IEEE 14th International Conference on Nanotechnology (IEEE-NANO), pp. 1000–1003. IEEE (2014)
3. Li, W., Birdwell, A.G., Amani, M., Burke, R.A., Ling, X., Lee, Y.H., Liang, X.L., Peng, L.M., Richer, C.A., Kong, J., Gundlach, D.J., Nguyen, N.V.: Broadband optical properties of large-area monolayer CVD molybdenum disulfide. Phys. Rev. B **90**(19), 195434 (2014)
4. Peng, T., Jian-Jian, X., Chao, Z., Shi, W., Run-Feng, C.: Graphene-like molybdenum disulfide and its application in optoelectronic devices. Acta Phys. Chim. Sin. **29**(04), 13–23 (2013)
5. Dodabalapur, A., Akinwande, D., Ha, T.J., Lee, J.: Method to improve performance characteristics of transistors comprising graphene and other two-dimensional materials. U.S. Patent Application No. 14/454422 (2015)
6. Kang, N., Paudel, H.P., Leuenberger, M.N., Tetard, L., Khondaker, S.I.: Photoluminescence quenching in single-layer MoS_2 via oxygen plasma treatment. J. Phys. Chem. C **118**(36), 21258–21263 (2015)

Microcontroller-Based Coffee Acidity Detection Design

Chao Zhang, Tao Jiang, Yonghang Tai, Zhikun Yang, Weimin Mao, and Huaiyi Zhang

Abstract In this paper, a coffee acidity detector with a coffee acidity sensor, an AT89C51 microcontroller, and an A/D converter is designed to detect coffee acidity, and it also has an audible and visual alarm function. This detector can not only detect the coffee acidity in a specific environment, but also set different thresholds by different environments, and give an audible and visual alarm when the coffee acidity in the air exceeds the set threshold value. The design is divided into two parts: Software design and hardware design. The hardware design part is to use the MQ303A coffee acidity sensor to first measure the coffee acidity in the air, and then convert it into a voltage signal, and then convert it into a digital signal with an A/D converter, and then transmit it to the microcontroller system. Finally, the microcontroller system and the corresponding peripheral circuit to perform signal processing, liquid-crystal display (LCD) if the caffeic acid value exceeds the threshold will produce an alarm. The software design part of the program adopts a modular design approach so that the functions of each subroutine are relatively independent and easy to debug and change. The circuit can be divided into a minimum system circuit for the microcontroller, an audible and visual alarm circuit, an A/D converter circuit, a keypad circuit, and an LCD circuit.

1 Introduction

Coffee is involved in several sub-sectors and processes, such as manufacturing, planting, testing, and so on. The key coffee-growing regions of the world in this region are Africa, the Americas, and India, and the Pacific. Based on export volumes from coffee-growing regions, we see that world demand for coffee is supporting a

C. Zhang · T. Jiang (✉) · Y. Tai (✉) · Z. Yang · W. Mao · H. Zhang
Yunnan Key Laboratory of Optoelectronic Information Technology, Yunnan Normal University, Kunming, China
e-mail: jiang_jt@126.com

Y. Tai
e-mail: taiyonghang@ynnu.edu.cn

steady growth, which has been relatively fast since 2005. This coincides with the growing awareness of coffee and the rapid growth of major potential markets around the world. The import and export of coffee are intertwined with the consumption situation, with Europe and the United States being the largest concentrated consumption regions, and coffee as the main local beverage has great viscosity, and in these countries and regions such as Europe, the United States, Russia, and Japan in Asia, coffee has a large consumer market and continues to maintain the growth of coffee consumption. The demand for coffee varies from region to region. From the per capita point of view, Europe and the United States have a great advantage, Finland's highest about 13 kg per person per year, the other side of the ocean in the United States about 4.3 kg, for coffee, in China is an imported product, the development history is short, about 0.04 kg per capita, and these regions compared to a big gap, but the people slowly accept the coffee culture, in the domestic development space day and night to enhance.

The design can be better and easier for everyone to know the approximate acidity of coffee by testing the acidity value to control the roasting industry, determine the type of coffee, increase the stability of the product, increase sales, to provide consumers with a cup of sweetness, acidity, bitterness, mellow, clean, and smooth coffee. In the process of coffee production, the food quality and safety inspection and testing process, as well as compliance with the SCAA standards and food safety, the use of coffee acidity testers, together with the correct testing methods, to obtain objective and accurate data.

2 Basic Model

2.1 Main Circuit

In the hardware design process, since the sensor can convert the non-electric energy of coffee acidity into electrical energy, it can output a voltage value of 0–5 V, which is relatively stable and the external interference is negligible. Therefore, the output voltage value of the sensor can be sent directly to the microcontroller unit (MCU) for processing through the ADC0832 data acquisition data [1, 2]. The hardware circuit design of the coffee acidity detection system is mainly: AT89C51 microcontroller system, sensor measurement circuit, LCD circuit, A/D conversion circuit, sound and light alarm circuit, and keyboard circuit. The circuit block diagram of the coffee acidity detection system hardware design is shown in Fig. 1.

The control module is mainly done by a microcontroller. The microcontroller controls the coffee acidity sensor, LCD, sound and light alarms, a series of data transmission, and finally realize the alarm function. The LCD is chosen to reach the display module [3]. Because it meets the requirements of the design, the P0 port of the microcontroller and LCD can be connected to transmit the data processed by the

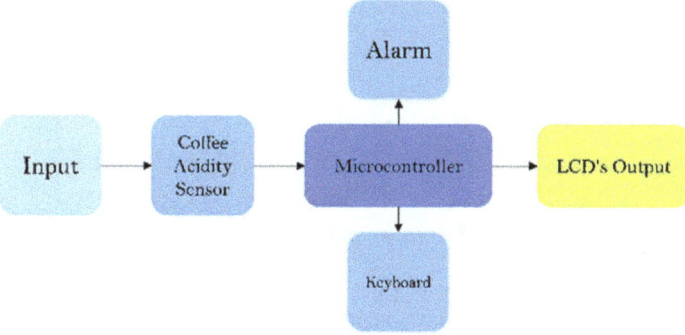

Fig. 1 Overall block diagram of the hardware program

microcontroller, so that the coffee acidity value on the LCD. Through a series of programming, the desired result can be achieved by the design.

2.2 AT89C51

A microcontroller is an integrated circuit chip, also known as a microcomputer, which is small, but five-fold computer system. These circuits can be controlled by software to complete tasks quickly, accurately, and efficiently [4]. In this design, we choose the microcontroller AT89C5, the core of the control, the key is based on the consideration of the AT89C51 is not decrypt low-power, ultra-low-cost and fast, high reliability, anti-static ability, anti-interference ability, more powerful microcontroller efficacy [5]. The AT89C51 has a total of 114 general-purpose input and output ports [6]. Four 16-bit timers, three 12-bit analog-to-digital converters, up to thirteen communication interfaces, and a DMA containing twelve independent transmission channels are available. As can be seen from the above characteristics, both the processor speed, the number of built-in memory components, and the various communication interfaces can meet the design requirements.

The on-chip structure of the AT89C51 microcontroller is shown in Fig. 2. It has all those basic elements necessary for manipulating the application on a fixed-sized integrated circuit chip [7]. According to the division of its efficacy, it is composed of the following components.

All of the above functional components are connected to a bus within the micro-controller, which is still based on the traditional construction method of the CPU and then the peripheral chip. However, the CPU uses the centralized control method of special function registers [8] for the control of each function part.

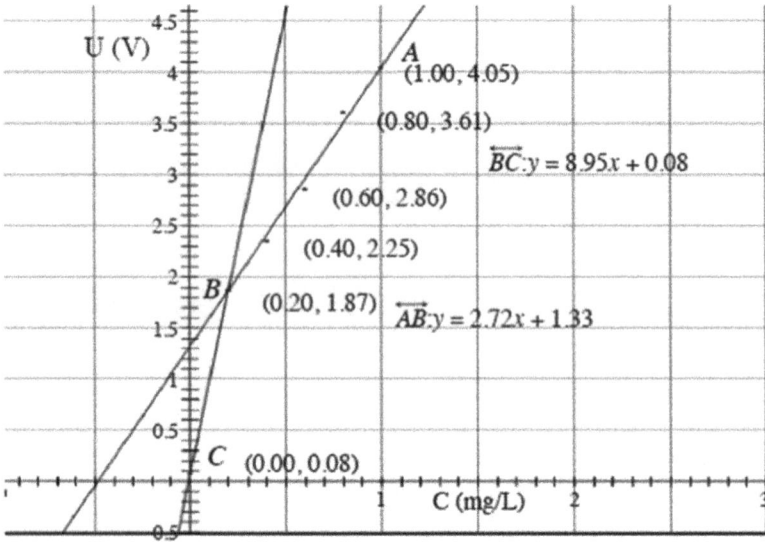

Fig. 2 The relationship between coffee acidity and output voltage

2.3   Coffee Acidity Sensor

The MQ303A coffee acidity sensor is a gas-sensitive sensor, compared to other sensors, it has a long life, high sensitivity, and stability of these advantages [9]. MQ303A coffee acidity sensor is fixed in stainless steel or plastic cavity inside by micro Al_2O_3 ceramic tube, SnO_2 sensitive layer, and the heater and measuring electrodes constitute a sensitive element. The relationship between coffee acidity and the output voltage is shown in Fig. 4, MQ303A coffee acidity sensor module is shown in Fig. 4. To make the measurement accuracy can reach the highest, the error is also the smallest, in the measurement of measurement before the sensor must be preheated for the 20 s. In the software simulation, because of the limited conditions, I used the voltage of the two ends of the sliding varistor to replace the output voltage of the coffee acidity sensor [10]. The relationship between coffee acidity and output voltage is show in Fig. 2.

The LCD with LCD acidity value is based on this function, and the accuracy of the algorithm is good.

3   Experiment

The main program achieves all the functions of a portable coffee acidity meter in conjunction with the hardware. Keys are data storage, display and detection, and the use of function sub-functions. The main program flowchart is shown in Fig. 3.

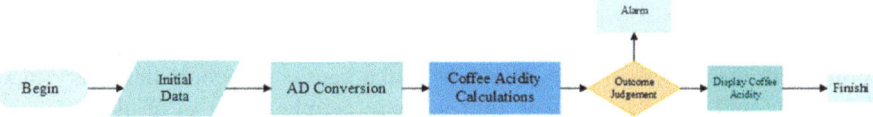

Fig. 3 Main program block diagram

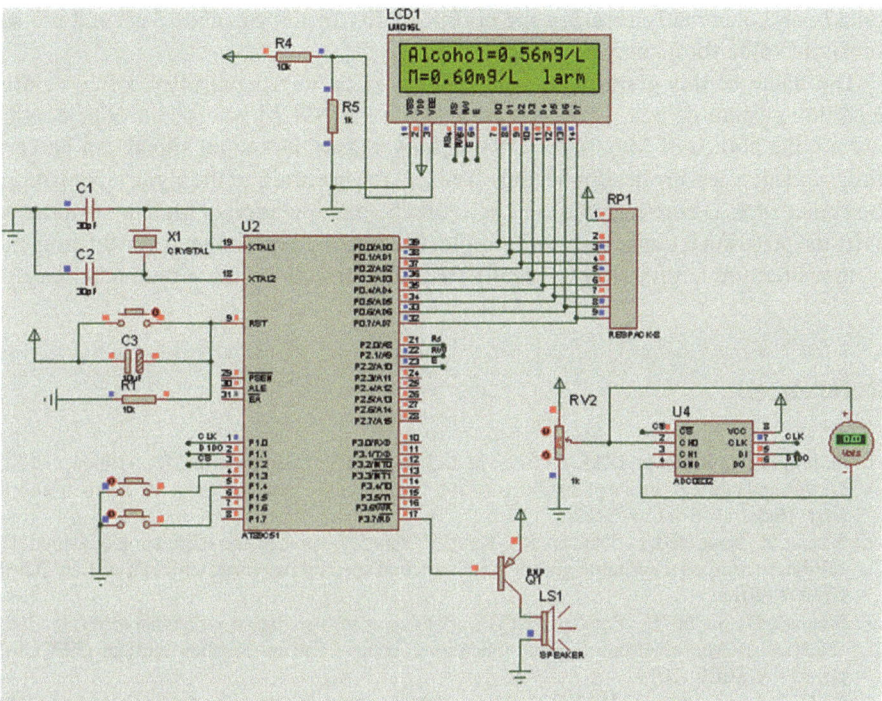

Fig. 4 Simulation results show

After the ADC0832 is initialized, it converts the zero to five-volt analog signal from channel 0 into the corresponding digital OOH-FFH and then stores the corresponding value into the memory unit, OOH-FFH is a hexadecimal number. The threshold for setting the alarm is 0.60 mg/L. When the coffee acidity in the air is detected as 0.56 mg/L, the coffee acidity in the air is less than the set threshold, the system does not alarm and the buzzer does not emit sound. The simulation results are as follows (Fig. 4).

4 Conclusion

Along with the growth and development of the economy, people have very much knowledge and contact with new things, more people began to drink coffee and understand coffee, so the coffee acidity detection in production, life, and other aspects have an important impact. For people in industrial production and daily life, the requirements of the coffee acidity detector, mainly to ensure that the detector can quickly and accurately measure the coffee acidity of the gas and to exceed the set threshold value for display and alarm.

The name of this graduation design is based on microcontroller coffee acidity detection system design, is an according to the AT89C51 microcontroller control system, the choice of MQ303A coffee acidity sensor, its output signal can be very direct and microcontroller connected, to ensure the accuracy of the signal acquisition; the choice of A/D converter and LCD circuit, the manipulation of high precision, and write the program is convenient and simple, through the experimental verification, the system structure is very simple, small size, rapid detection, has certain practicality.

References

1. de Jong, B.R., Brouwer, D.M., de Boer, M.J., Jansen, H.V., Soemers, H.M.R., Krijnen, G.J.M.: Design and fabrication of a planar three-DOFs MEMS-based manipulator. J. Microelectromech. Syst. **19**(5), 1116–1130 (2010)
2. Xi-Lin, Z., Song-Tao, L., Dan, L., Rui-Kang, Z., Chen, J.: Design and fabrication of a 400 GHz InP-based arrayed waveguide grating with flattened spectral response. Chin. Phys. Lett. **32**(5), 054202 (2015)
3. Santoso, D., Susilo, D., Prasetyanto, Y.Y.: A microcontroller-based yarn twist tester. In: 2014 Electrical power, electronics, communications, control and informatics seminar (EECCIS), pp. 35–39. IEEE (2014)
4. Li, G.Z., Liu, C., Cao, G.H., Yu, N.: Design for detection system of landslides based on SCM. Appl. Mechan. Mater. Trans. Tech. Publ. Ltd. **331**, 622–625 (2013)
5. Minhua, W., Zhixiong, H., Yuanlong, W.: Design of detection system for fault of train angle cock based on AT89C51. Comput. Measur. Control, 42–43+56 (2008)
6. Xiong, Z.G., Jiang, P., Hu, W.W., Luo, Y.H., Peng, K.: Design of STC microcontroller-based intelligent remote water tower cluster monitoring system. Hubei Agric. Sci. **52**(14), 3415–3419 (2013)
7. Chahroudi, D.: Solar Control System. US (1981)
8. Li, J., Zhang, W., Xu, B.: Design of two-wheeled self-balancing robot control system. Adv. Mater. Res. Trans. Tech. Publ. Ltd. 1044–1045, 774–777 (2014)
9. Minsky, M.L.: Computation: finite and infinite machines. Am. Math. Mon. **75**(4), 428–429 (1968)
10. Khouni, L., Khellaf, A., Saoud, L.S., Mokeddem, D.: A microcontroller based system for an automatic control of ph and acidity degree of chemical products. Int. Rev. Comput. Softw. (2007)

Experimental Study on Measuring Acceleration of Gravity with Simple Pendulum

Licun Sun, Chao Zhang, Yonghang Tai, Zhikun Yang, and Baocheng Dong

Abstract The vibration of a single pendulum is considered as a rigid body rotation, and the effects of pendulum angle, pendulum mass distribution, air buoyancy and system damping on the vibration of a single pendulum are considered. At the same time, the experimental device of single pendulum is slightly improved by adding a dial that can read the angle at the top of the single pendulum instrument to facilitate the control of the pendulum angle, and the video processing software is used to analyze the motion video of the single pendulum for the purpose of accurate period measurement.

1 Introduction

The single pendulum is a classical model in vibration and is assigned to be presented in the section on simple harmonic vibration. The single pendulum experiment to measure the acceleration of gravity is an important one, covered in both college and high school. A small ball with a non-extendable light rope hanging, the upper end of the rope fixed at one point, keeping the rope straight, pulling the ball and the vertical direction of the deviation from a certain angle θ, no initial velocity release of the ball, so that it is in a plane to do amplitude in a small swing is a single pendulum.

Due to the buoyancy of the air, the mass distribution of the pendulum, the existence of system damping, and the incorrect operation of the experimenter, errors will occur in the measurement of gravitational acceleration. The experimental error is reduced from the following two aspects: First, correct the measurement formula of the acceleration of gravity to reduce the theoretical error; second, improve the simple pendulum experimental device to reduce the use of the instrument in the experimental

L. Sun · C. Zhang (✉) · Y. Tai (✉) · Z. Yang · B. Dong
Yunnan Key Laboratory of Optoelectronic Information Technology, Yunnan Normal University, Kunming, China
e-mail: zhchynnu@foxmail.com

Y. Tai
e-mail: taiyonghang@126.com

operation. The error generated therein. In this paper, we will combine the above two methods to improve the single-pendulum experiment, and also combine the computer graphics processing, and strive to use the minimum cost to get the best results [1], while increasing students' interest in the experiment to achieve the teaching purposes of physics experiments.

2 Basic Principle

2.1 Principle of Single Pendulum Experiment

Suppose the ball is swinging with an amplitude angle of θ, the mass of the ball is m, the distance from the suspension point to the center of mass of the ball is l. The ball is subjected to the pulling force of the rope and its own gravity, which decomposes along and perpendicular to the direction of the rope, along which the ball is balanced by the forces in the rope. The magnitude of the force in the tangential direction of the rope is $mg \sin \theta$, such that when the angle is small, the magnitude of the tangential force is approximately equal to $mg\theta$[2]. Equations (1) and (2) show the equations of motion of the ball. To satisfy the simple harmonic equation of motion [3], the expression for tension is as in Eq. (3).

$$\mathrm{ml}\frac{\mathrm{d}^2\theta}{\mathrm{d}t^2} = -mg\theta \tag{1}$$

$$\frac{\mathrm{d}^2\theta}{\mathrm{d}t^2} = -\frac{g}{l}\theta \tag{2}$$

$$T = \frac{2\pi}{w} = 2\pi\sqrt{\frac{L}{g}} \tag{3}$$

2.2 Correction of the Principle

Although they are all convolutional neural models, dimensionality reduction using the one-dimensional or K-max pooling method and the most obvious features indicate different effects. Finally, pass it to the fully connected layer to get the output result. The formula for measuring the acceleration of a single pendulum in university physics experiments, which is an approximate formula that does not take into account the effect of pendulum angle size, pendulum mass distribution, air buoyancy, and other factors on the vibration of a single pendulum [4], and the formula for calculating the acceleration of gravity must be modified to take these factors into account.

When a single pendulum makes an oscillation with a small deflection angle θ, consider the vibration of the single pendulum as the rotation of a rigid body, according to the law of rotation (4). The pendulum ball's rotational inertia, access to the integral table, the period of motion of the pendulum ball as shown in Formula (5), the acceleration of gravity relation in accordance with (6).

$$mgl \sin \theta = -I \frac{d^2\theta}{dt^2} \tag{4}$$

$$T = 2\pi \sqrt{\frac{l}{g}} \left[1 + \left(\frac{1}{2}\right)^2 \sin^2 \frac{\theta}{2} + \left(\frac{1}{2} \times \frac{3}{4}\right)^2 \sin^4 \frac{\theta}{2} + \cdots \right] \tag{5}$$

$$mgl \sin \theta + \mu g \frac{l}{2} \sin \theta = -\left(ml^2 + \frac{2}{5}md^2 + \frac{1}{3}\mu l^2 \right) \frac{d^2\theta}{dt^2} \tag{6}$$

Taking the above into account, the overall correction for the acceleration of gravity by a simple pendulum can be written as:

$$\varepsilon = \left(1 + \frac{1}{2} \sin^2 \frac{\theta}{2} + \frac{2}{5} \frac{d^2}{l^2} - \frac{1}{6} \frac{\mu}{m} + \frac{\rho_0}{\rho} \right) \tag{7}$$

$$g = 4e\pi^2 \frac{l}{T^2}(1 + \varepsilon) \tag{8}$$

In the above discussion, the effect of air resistance and self-damping is ignored. In the actual vibration process, due to the effect of these resistance [3], $\sin \theta = 0$, the motion formula of simple pendulum can be expressed as $\theta^2 + 2\beta \dot{\theta} + \omega_0^2 \theta = 0$, this is a second-order linear ordinary differential equation. Under the initial state conditions of $\theta(0) = 0$ and $\dot{\theta}(0) = 0$, the following results can be obtained:

$$\theta(t) = \frac{\theta_0}{\sqrt{\beta^2 - \omega_0^2}} e^{-\beta t} \left(\beta \frac{-e^{-t}\sqrt{\beta^2 - \omega_0^2} + e^t \sqrt{\beta^2 - \omega_0^2}}{2} + \sqrt{\beta^2 - \omega_0^2} \frac{-e^{-t}\sqrt{\beta^2 - \omega_0^2} + e^t \sqrt{\beta^2 - \omega_0^2}}{2} \right) \tag{9}$$

$$\theta(t) = \theta_0 e^{-\beta t} \left(\cos \sqrt{\beta^2 - \omega_0^2} + \frac{\beta}{\sqrt{\beta^2 - \omega_0^2}} \sin \sqrt{\beta^2 - \omega_0^2} t \right) \tag{10}$$

In the actual vibration, the simple pendulum is an underdamped vibration, and its circular frequency is $\omega = \sqrt{\omega_0^2 - \beta^2}$. Here, β can be measured according to the attenuation of amplitude. If the initial amplitude is, θ_0 the amplitude changes to θ_n after n oscillations, assuming that the amplitude attenuation of a simple pendulum becomes general after N cycles of vibration, the expression of ω can be obtained as follows

$$\beta = \frac{\omega}{2\pi n} \ln \frac{\theta_0}{\theta_n} \tag{11}$$

$$\omega = \frac{\omega_0}{\sqrt{1 + \left(\frac{\ln(\theta_0/\theta_n)}{2\pi n}\right)^2}} \tag{12}$$

$$\omega = \frac{\omega_0}{\sqrt{1 + \left(\frac{\ln 2}{2\pi n}\right)^2}} \approx \omega_0 \left(1 - \frac{0.006085}{n^2}\right) \tag{13}$$

In the actual situation of simple pendulum vibration, the length of cycloid will change in the process of swing because of its elasticity, but this factor has very weak influence on the measurement result of gravity acceleration, so it is not corrected. In the actual operation process, if the pendulum ball is accidentally given an initial velocity, it may change the swing of the ball into a cone pendulum, and also cause errors in the measurement results of the experiment.

3 Experiment

3.1 Simple Pendulum Experimental Device and Measurement Method

In recent years, physical experiments toward the development of intelligent direction [5], the use of electronic components combined with computer data acquisition, so that experimental operations are more accurate and more convenient. In the actual measurement process, the measurement error of pendulum length is large. In the related literature, the measurement method similar to Vernier Caliper is used to improve the measurement of pendulum length [6]. When one end of the cycloid mark is pressed on the tape measure, a small object can be placed under it to make the cycloid parallel to the plane of the tape, so as to further improve the measurement accuracy. The electronic stop watch is used in laboratory to measure the period of simple pendulum. Its precision is 0.1 s. The reaction time of human according to the table may lead to the deviation of measurement result [7]. Therefore, it is necessary to keep the camera or mobile phone fixed when recording video.

The process of pendulum video processing with tracker software is as follows: (1) Import video to establish scale. Find an object of known size in the video as a reference, calibrate it with a scale and set its real length; (2) Establish a coordinate system. Click the coordinate button to establish the coordinate system, and drag the origin of the coordinate to the balance position of the pendulum. If the video direction is inclined, the coordinate axis can be rotated to make the Y coordinate axis coincide with the cycloid of the pendulum when balancing; (3) Establish a particle to track the position of the swing ball. Right click the create icon to create particle A. hold

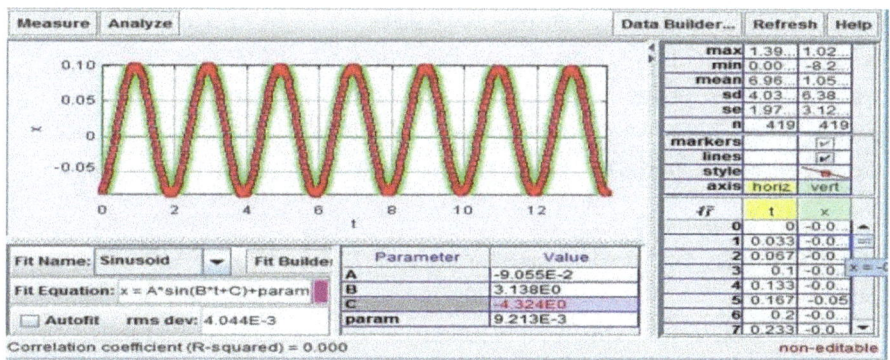

Fig. 1 Data fitting with tracker

down the shift key and right-click the mouse to calibrate the position of the swing ball in each frame of image. You can also press and hold the shift and Ctrl keys at the same time. After the circular selection area appears, click the position of the swing ball and zoom the size of the video, so that the circular selection area can select the swing ball exactly, and then select automatic tracking to collect data to close the dialog box. (4) Data analysis, operation as, shown in Fig. 1.

3.2 Experimental Results and Analysis

Using the modified formula and the improved method, the data of the experiment of measuring the acceleration of gravity with a pendulum are collected, and the value of the acceleration of gravity is calculated. The results and average values of the five measurements are shown in Table 1.

The mass of the ball $m = 30.16$ g, the mass of one-meter cycloid $\mu = 0.36$ g. According to the measured data, the density of the ball is $\rho = 7.205$ g/cm^3, the air density is $\rho_0 = 1.29 \times 10^{-3}$ g/cm^3, and the gravity acceleration in Kunming area is 9.783 m/s^2. The pendulum is allowed to swing at different angles from the vertical direction to measure the acceleration. The deflection angles are $3°, 5°, 10°, 15°, 20°, 25°, 30°, 35°$ and $40°$, respectively. The measurement results are shown in Tables 2 and 3.

The gravity acceleration is measured by single pendulum with pendulum length of 60, 70, 80, 90, 100, 110 and 120 cm, respectively. The measurement results are shown in Table 4.

Table 1 Measurement pendulum diameter

Frequency	1	2	3	4	5	Average value
d/cm	1.990	2.002	2.002	1.994	2.008	1.992

Table 2 Gravity acceleration with different swing angles

Deflection angle	3°	5°	10°	15°	20°
T/S	1.995	1.9974	2.0006	2.0018	2.0106
g/m/s²	9.830682	9.807683	9.779186	9.772212	9.69344
δ relative error	0.004874	0.002523	0.00039	0.0011	0.00915

Table 3 Gravity acceleration with different swing angles

Deflection angle	25°	30°	35°	40°
T/S	2.0126	2.0188	2.028	2.0324
g/m/s²	9.682622	9.633477	9.558241	9.530627
δ relative error	0.01026	0.01528	0.02297	0.0258

Table 4 Gravity acceleration measurement with different pendulum

Deflection angle	80	90	100	110	120
T/S	1.7862	1.8982	1.995	2.1	2.1918
g/m/s²	7.849255	8.796034	9.830682	10.73505	11.7279
δ relative error	0.19766	0.10089	0.004874	0.097317	0.19877

From the above results, it follows that the error is the smallest when the pendulum length is 100 cm.

4 Conclusions

In this experiment, the period formula of simple pendulum is modified, and the correction items of swing angle, cycloid mass distribution, air buoyancy and system damping are introduced. Combined with the improvement of experimental device and measurement method, the experiment of measuring gravity acceleration with simple pendulum is improved. By analyzing the measured data, it is found that the deflection angle of simple pendulum is within 15°, and the pendulum length is about 1 m. The accuracy of gravity acceleration can be improved by using the improved method.

References

1. Jin-de, W.: Experimental study and error analysis in the experiment of "measuring acceleration of gravity by single pendulum." J. Tongren Univ. **13**(03), 132–134 (2011)

2. Frémond, M., Shitikova, M.V.: Non-smooth thermomechanics. Appl. Mech. **55**(5), B99–B100 (2002)
3. Kresse, G., Hafner, J.: Ab initio molecular dynamics for liquid metals. Phys. Rev. B **47**(1), 558 (1993)
4. Zhong-zhi, Y.D.W.S.: More discussion on maximum pendulum angle of measuring acceleration of gravity with simple pendulum. Phys. Exp. **7**, 44–45+48 (2003)
5. Amin, M.A., Sobhani, Z., Chadalavada, S., Naidu, R., Fang, C.: Smartphone-based/fluoro-SPE for selective detection of PFAS at ppb level. Environ. Technol. Innov. **18**, 100778 (2020)
6. Ertas, A., Mustafa, G.: Real-time response of the simple pendulum—an experimental technique. Exp. Tech. **16**(4), 33–35 (1992)
7. Yong, Z., Geng-lei, L., Xiao-ping, Z.: Measuring the period of simple pendulum using force sensor (10), 26–28 (2008)

Design of Intelligent Boiler Controller

Sijia Lian, Tao Jiang, Chao Zhang, Zhikun Yang, and Shaoquan Jiang

Abstract This paper discusses the development status of boilers in China. The design and implementation of the intelligent boiler controller are introduced in detail. The author completed the controller using the STC89C51 microcontroller, temperature sensors, liquid crystal displays, electromagnetic relays, waterproof probes, water level sensors, and other components. The intelligent boiler controller has the advantages of simple structure, small size, and convenient operation. It can adjust the target temperature according to different requirements and display the real-time temperature.

1 Introduction

With the continuous progress of science and technology and the continuous improvement of living standards, people's demand for intelligent products continues to rise. Product intelligence [1] has become a hot research direction. For example, the extensive use of sweeping robots shows that the development speed and popularity of smart small household appliances have exceeded people's imagination. Intelligent improvement of small household appliances has also become a trend [2]. In addition, with people's demand and yearning for a convenient life, ordinary household water heaters began to be favored by consumers. Among them, the ordinary small boiler only has the function of heating water to boiling, but cannot control and adjust the temperature, and its function is very single. In modern life, especially in the office and other working environments, people want to drink a cup of water with proper temperature in time, but the water burned by the conventional boiler has reached the boiling point. In such a big environment, this design is to realize the intelligent design of the boiler, using STC89C51 single-chip microcomputer [3, 4] to add temperature

S. Lian · T. Jiang (✉) · C. Zhang (✉) · Z. Yang · S. Jiang
School of Physics and Electronic Information, Yunnan Normal University, Kunming, China
e-mail: jiang_jt@126.com

C. Zhang
e-mail: zhchynnu@foxmail.com

© The Author(s), under exclusive license to Springer Nature Singapore Pte Ltd. 2021   275
L. C. Jain et al. (eds.), *3D Imaging Technologies—Multidimensional Signal Processing and Deep Learning*, Smart Innovation, Systems and Technologies 236,
https://doi.org/10.1007/978-981-16-3180-1_34

control function, which can set the temperature according to different requirements, for example, adjust the target temperature to 50–55 °C by adding and subtracting buttons.

Intelligent design and improvement of water boiler can add new functions to its traditional mode, and provide more services and greater convenience for users. It is believed that the intelligent boiler controller will have a place in the future home furnishing industry.

2 Hardware Schematic Diagram

2.1 General Schematic Diagram of Circuit

In this design, the intelligent boiling water controller circuit uses STC89C51 micro-controller as microcontroller, and adds power supply module, indicator lamp module, electromagnetic relay module, liquid crystal display module, buzzer alarm module, temperature measurement module, water level detection module, key module, temperature control module and remote-control receiving module. The hardware schematic diagram of the intelligent boiling water controller designed this time is shown in Fig. 1.

2.2 PCB Wiring Diagram

Based on Altium designer 10 software, this design draws power supply module, indicator lamp module, electromagnetic relay module, liquid crystal display module, buzzer alarm module, temperature measurement module, water level detection module, key module, temperature control module circuit module, etc. into the schematic diagram for adjustment, modification, and planning, and uses reasonable components to package the design. By adjusting the position of each component and the line width between each component, the layout of the circuit board is more concise and reasonable. The final wiring diagram is small in size, simple in structure, and reasonable in layout, which meets the requirements of this design.

2.3 Copper Clad Laminate

According to the characteristics of intelligent boiler controller, adopt reasonable components and weld each component. Copper-clad laminate (CCL) is selected for the soldered circuit board of this design. Copper-clad laminate is made by dipping the material into resin, covering one side with copper foil and hot pressing. By then, the

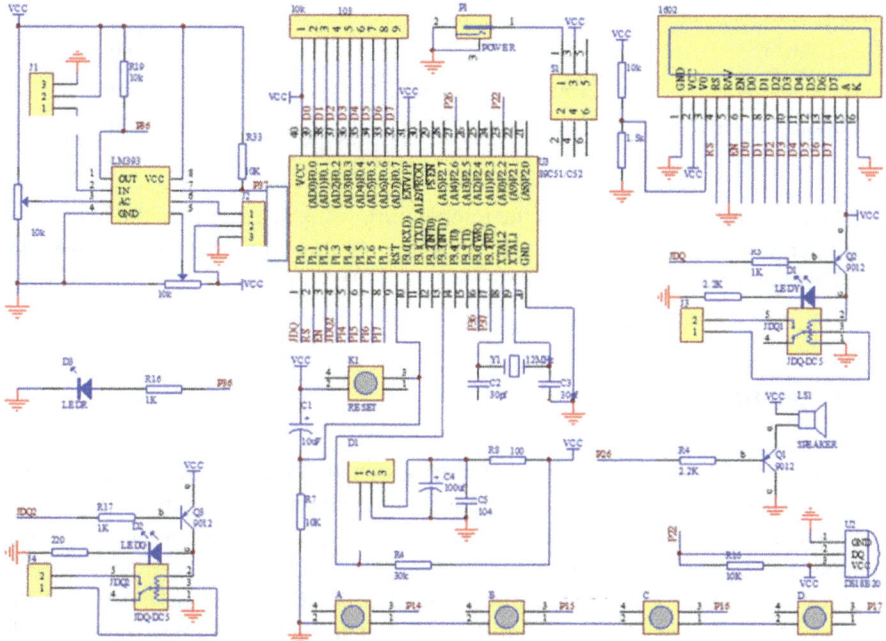

Fig. 1 Complete schematic diagram

circuit corrosion and other processes have been completed, which makes the design more convenient and simpler, omits the complicated connection process of circuit components, and thus avoids the poor contact of wires and the inability to connect wires when the circuit is connected. And the copper-clad plate has the advantages of low working temperature, strong moisture resistance, strong heat resistance, and long service life, so it is the best choice for soldering circuit boards in this design.

3 Software Construction of Intelligent Boiling Water Controller

The main working path of the software program design of this design is: after the program starts, initialize the setting parameters, that is, the user inputs the target temperature. Then enter the cycle, during which the temperature is first read and then displayed on the display screen. Then judge the water level, and if the water level is lower than the drainage line, start heating; If it is higher than the upper water level, stop heating; If there is an error state, that is, higher than the upper water line and lower than the lower water line, the alarm will continue until the error is manually released. After judging the water level state, judge the temperature value [5]. If the temperature value is higher than the upper limit of the set temperature,

Table 1 Data for actual temperature

Thresholds/°C	50	60	70	80
Final temperature/°C	50.3	60.2	70.1	80.2
Steady-state errors/°C	0.3	0.2	0.1	0.2

stop heating and give an alarm; If the temperature value is lower than the lower limit of temperature, start heating and give an alarm. At this point, the cycle is over, and the program is over.

4 Experiment

In order to confirm the accuracy of the designed boiling water temperature, a large number of temperature measurement experiments were carried out on the water reaching the target temperature, and several representative temperatures were summarized, as shown in Table 1.

According to the measured temperature data above, it can be concluded that the design meets the expected requirement that the temperature error should be guaranteed within 1 °C. In addition, the precision of the numerical value displayed on the display screen is kept as one decimal place [6].

5 Conclusion

Because of the use of electromagnetic relay, the intelligent water boiling controller designed this time does not need to consider the problem of switch reset at all; Real-time temperature can be displayed on the display screen, as well as the target temperature upper limit and temperature lower limit set by the user; The problem of closing the lid of the kettle and the volume of water has also been solved; In addition, the unnecessary waste caused by the large temperature difference in the fixed default boiling mode technology is perfectly avoided; It meets the design requirements and has a good market application prospect. It makes up for the defects of ordinary household water heaters, thoroughly optimizes the shortcomings and deficiencies in the discussion, realizes the expansion of functions, truly realizes the intelligentization of small household appliances, and gives users no worries, and implements the sustainable development concept of energy saving.

References

1. Barat, V., Borodin, Y.R.I.J., Kuzmin, A.L.E.X.E.Y.: Intelligent AE signal filtering methods. J. Acoust. Emission **28**, 66–77 (2010)
2. Xu-Wu, S.U., Zuo, F.: Research on the generation of the disassembly sequence of small-size waste household appliances based on modularization. In: 2016 International Conference on Artificial Intelligence: Techniques and Applications, p. 6. Shanghai (2016)
3. Li, W., Li, Y., Fan, X.: The design and implementation of digital temperature measurement and automatic control system. In: 2010 International Conference on Computer Application and System Modeling (ICCASM 2010), pp. v10407–v10409. IEEE (2010)
4. Li, W., Chen, Z.P., Peng, X.R., Zhu, H.Y., Wang, T.: Exploration of teaching reform of single chip microcomputer application technology course under the cooperation of school and enterprise. Int. J. Comput. Eng. **4**(4), 217–220 (2019)
5. Pereira, T.C., Lopes, R.A., Sciubba, E.: Exploring the energy flexibility of electric water heaters. Energies **13**(1), 46 (2019)
6. Chen, M.F., Hu, X.D.: Design of temperature control system based on AT89S51 single-chip microcomputer. Mech. Eng. **01**, 141–142 (2009)

Design of Fire Alarm Based on 51 Single Chip Microcomputer

Yuexiong Feng, Zhikun Yang, Chao Zhang, Tao Jiang, and Shaoquan Jiang

Abstract In this paper, 51 single-chip microcomputer is designed and implemented. In this design, 51 single-chip microcomputer is the main control chip, MQ-2 smoke sensor is used to detect the concentration of combustible gas and combustion smoke in the air in real-time, and DS18B20 temperature sensor is used to monitor the temperature of the environment. The above data is displayed on LCD1602 in real-time, or the TTL level signal sent by 51 single-chip microcomputer is converted into RS232 signal through MAX232 chip, and finally uploaded to the safety control hub of the building, so as to realize the input monitoring of whether there are hidden dangers in all parts of the building. When there is a hidden danger in the building, the system will automatically send out an alarm through the sound and light alarm circuit, and the personnel can also press the alarm button to trigger the alarm. Through the modular design, the system establishes a complete monitoring and alarm system by building reset module, clock module, key detection circuit, sound and light alarm circuit, A/ D acquisition circuit, smoke sensor module, temperature sensor module, display module, etc.

1 Introduction

For building equipment, the requirements for fire safety are very high. Therefore, a set of safe and reliable building fire alarm system is very important for building equipment, which plays a key role in ensuring building safety and preventing alarm in time. The building system is complex and contains many parts. The space in the building is narrow, there are many combustibles, and the fire point is hidden and difficult to be found. Therefore, it is very necessary to design and manufacture a

Y. Feng · Z. Yang (✉) · C. Zhang (✉) · T. Jiang · S. Jiang
School of Physics and Electronic Information, Yunnan Normal University, Kunming, China
e-mail: ynkmyzk@126.com

C. Zhang
e-mail: zhchynnu@foxmail.com

© The Author(s), under exclusive license to Springer Nature Singapore Pte Ltd. 2021   281
L. C. Jain et al. (eds.), *3D Imaging Technologies—Multidimensional Signal Processing and Deep Learning*, Smart Innovation, Systems and Technologies 236,
https://doi.org/10.1007/978-981-16-3180-1_35

building fire alarm system which can accurately detect the fire in buildings and give out an alarm quickly.

At present, the fire alarm systems commonly installed in domestic building systems are mainly distributed intelligent and centralized intelligent. On this basis, some buildings with high requirements for safety monitoring have made corresponding technical improvements to upgrade their monitoring effect, but on the whole, it is based on decentralized intelligence and centralized intelligence. In the future, with the increasing demand for intelligent monitoring systems, full intelligent monitoring systems will become the main development direction and research focus. At present, China is focusing on the distributed intelligent fire monitoring and alarm system [1, 2]. The fire alarm system designed in this paper is mainly used in places with limited space. The fire monitoring sensor adopts MQ-2 smoke sensor and DS18B20 temperature sensor, through which the temperature and smoke are measured.

Data are monitored in real-time, and the monitoring signals are transmitted to the control unit in two ways. The above-monitored data can be displayed on LCD1602 in real-time and uploaded to the control center of the building through serial port. When the system detects that there is a fire or there is a fire hazard in the current environment, it will give an alarm, and when the personnel finds abnormal phenomena, they can also trigger the manual alarm by pressing the key.

2   Overall System Design

In the fire alarm system, the fire alarm controller is the control core of the whole system, which is usually installed in the bridge or fire control station. The alarm controller is used to receive the signals from the fire detector and the manual alarm button, and give an alarm in case of fire. There are two types of controllers: regional controller and fixed-point controller. Regional controller can identify the location of fire, while fixed-point controller can identify the specific location of each detector or alarm button. At present, regional fire alarm controllers are the main fire alarm systems used in buildings.

According to the requirements of relevant technical specifications, when the fire detector detects a fire, or the manual alarm button is pressed, the controller should respond quickly and send out sound and light signals to give an alarm. In addition, the indicating device should accurately display the area where the alarm detector or manual button is located. If the alarm is not confirmed within 2 min, the whole building public alarm will be triggered.

This paper is to study a building fire alarm system with single chip microcomputer as control device. Its main functions are: to monitor whether there is fire hazard in buildings in real time, display and upload the collected monitoring data, and realize the functions of automatic alarm and manual alarm for building fires.

3 Hardware Circuit Design of the System

3.1 51 Single Chip Microcomputer System

Miller hyperbolic function model: a mathematical model of hysteresis loop suitable [3]. In this scheme, 51 single chip [4, 5] microcomputer is used as the main control chip of the system, MQ-2 smoke sensor is used to collect A/D data to obtain the smoke concentration in the building, DS18B20 temperature sensor is used to detect the temperature of the space environment in the building, and the above parameters are displayed on LCD1602 in real time. The above parameters are uploaded to the building ship control center through RS232 serial port.

This system uses 51 as the control chip. The chip has many advantages, such as fast operation speed, easy control, rich functions, and so on, and it is small in size, high in integration and low in cost. Therefore, it is widely used in both scientific research and industrial control. The chip is equipped with 4 k bytes and 128 bytes, 32 ports, two timers, a full-duplex serial port, and an on-chip oscillator and clock circuit.

As shown in Fig. 1, AT89C51 is the smallest system that can work normally. The voltage VCC required for power supply of the system is 5 V, and the external crystal oscillator with Y1 of 12 M provides external clock input for the system. The manual reset circuit is designed in the system so that the staff can reset the system manually when the system program runs away.

3.2 Smoke Collection Circuit

See Fig. 2 for the power supply scheme of the system. The rated input voltage of the system is 12 V, which allows certain fluctuation. The voltage is converted by MP1584 switching power supply chip, and 5 V voltage is output to provide power for MCU. MP1584 is a high-frequency step-down chopper circuit controller, in which a high-power MOS transistor is integrated. The input voltage ranges from 4.5 to 28 V, and the highest working frequency can reach 1.5 MHz. It has the characteristics of wide input voltage range and high voltage conversion efficiency, thus reducing the heating of the system.

4 System Test

As shown in Fig. 3, it is the simulation diagram of automatic alarm triggered by high-temperature in building ships. It can be seen from the diagram that when the smoke concentration in the environment is low and the environment temperature is high, the simulation system sets the current environment temperature to 100 °C,

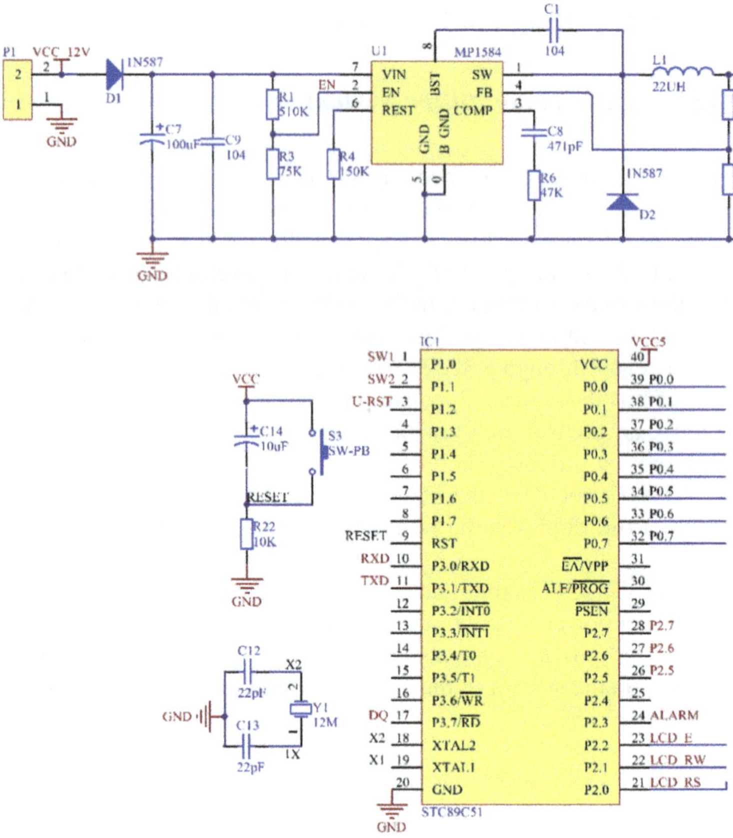

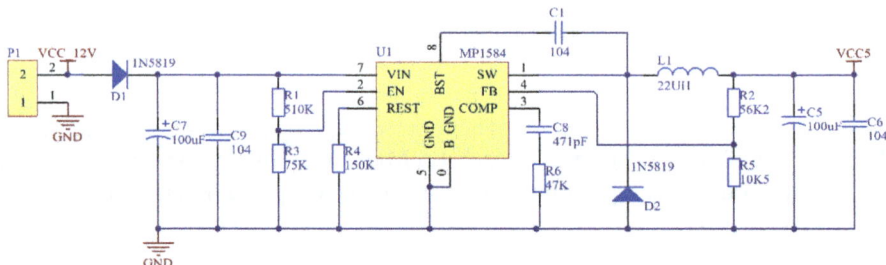

Fig. 1 Single chip microcomputer minimum system

Fig. 2 Schematic diagram of power supply for AT89C51 microcomputer system

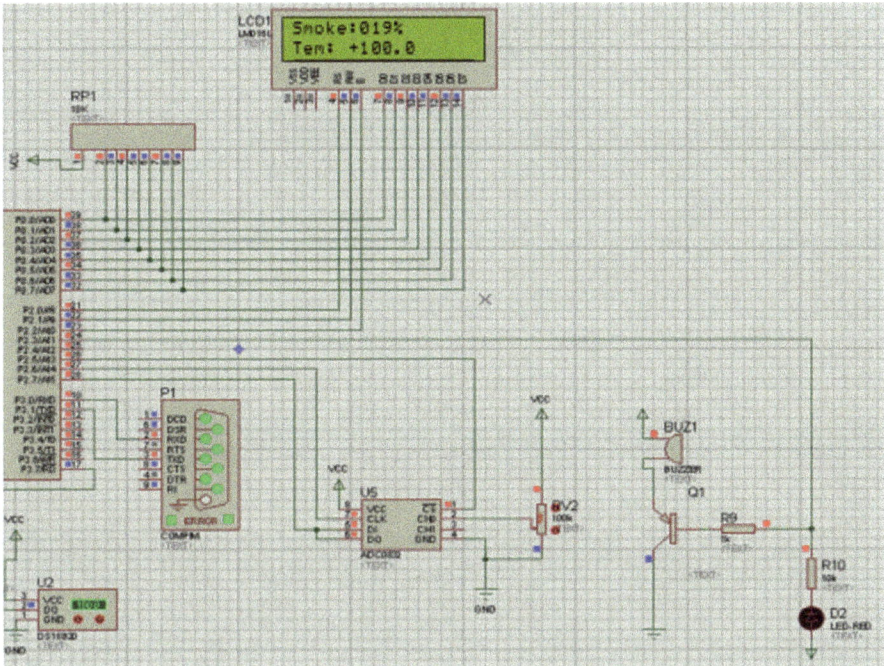

Fig. 3 Test diagram of high temperature trigger alarm function

which will trigger the automatic alarm of the fire alarm system. At this time, the buzzer alternately sounds and the LED lights flash.

Figure 4 shows the simulation diagram of the high smoke concentration in the building that triggers the automatic alarm of the alarm system. It can be seen from the figure that when the temperature of the environment is low, if the smoke concentration of the environment is high, the simulation at this time. The system sets the current ambient smoke concentration to 60%, which will trigger the fire alarm system to automatically alarm. At this time, the buzzer alternately buzzes and the LED lights flash.

5 Conclusion

In this paper, a convenient and efficient fire alarm system for buildings is designed. The 51 single chip microcomputer is used as the main control chip of the system, the clock circuit of the chip adopts external clock oscillation circuit, and the reset circuit of the chip adopts a reset mode combining power-on reset and manual reset. The MQ-2 smoke sensor module and DS18B20 temperature sensor are used to monitor the corresponding position of the building ship. The system uses LCD screen to

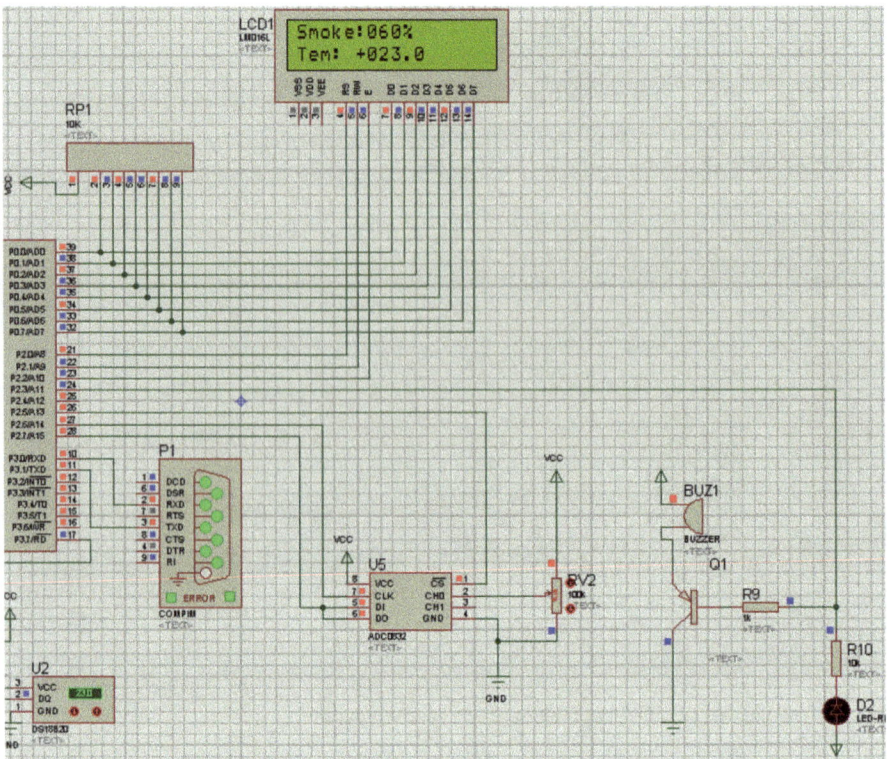

Fig. 4 High smoke concentration trigger alarm function test chart

display smoke concentration and ambient temperature in real time. When there is a
fire hazard in the building ship, the system can automatically detect it and trigger
the alarm equipment, i.e. buzzer and LED, to give an alarm to remind people to
extinguish the fire in time. At the same time, the system can trigger the alarm speed
of the fire alarm manually by pressing the key, or it can contact the fire alarm manually
by pressing the key. After testing, the system basically achieves the expected design
goals, has the function of fire alarm, and can work stably.

References

1. Xu, Y.L., Zhou, X.H., Cai, S.D., Xu, Y.C.: Research and development of gas sensors in china.
   Sens. Actuators, B Chem. **25**(1–3), 555–558 (1995)
2. Electronic, B.: Smart turbine automation erhoht wirtschaftlichkeit. Elektrotechnische Z. **138**(8),
   12–12 (2017)
3. Kernighan, B.W., Ritchie, D.M.: The C programming language—ANSI C. ACTUS, Software-
   Practice and Experience, vol. 25(1), 2nd edn, pp. 14–27 (1988)

4. Kuang, P., Cao, W.N., Liu, Z.G.: Design of dynamic screen system based on MCU. In: 2014 11th International Computer Conference on Wavelet Active Media Technology and Information Processing (ICCWAMTIP), pp. 468–470. IEEE (2014)
5. Qiu-hong, Y., Rui, T.: On the teaching reform of single chip microcomputer principle and application (07), 58–60 (2011)

Design of Multi-Input High Fidelity Preamplifier Based on STM32

Anlei Yu, Tao Jiang, Chao Zhang, Zhikun Yang, and Shaoquan Jiang

Abstract With the popularization of audio-visual entertainment equipment, the application of pre-stage power amplifiers is becoming more and more extensive. The use of a high-quality pre-stage amplifier will often bring people a better sound quality experience. The system uses STM32F103C8T6 as the main control chip, controls the multi-channel audio selector NJW1110 through the IIC bus, and can select the input audio signal corresponding to the output channel and the gain of the output audio signal. The output signal distortion is small, and HMI is used for human-computer interaction control The serial port screen greatly facilitates the user's operation and use.

1 Introduction

With the progress and development of science and technology, people's pursuit of high-fidelity power amplifier [1, 2] has been further improved. In order to control the high-fidelity power amplifier and better meet the needs of the public, the subject selects 9-input 3-output stereo audio selector NJW1110, which has excellent audio characteristics, low distortion, less than 0.0007%typ, low output noise, and low crosstalk. All internal states and variables can be controlled through IIC bus. In this project, STM32F103C8T6 is used as the control core to realize the control of input and output and gain control of power amplifier, so as to obtain higher quality at lower cost.

Inter integrated circuit bus is a high-performance serial synchronous communication transmission bus between chips. Compared with other serial buses of the same type, IIC bus is widely used in small batch data transmission and control. IIC bus simplifies the design of hardware circuit, and only serial data line (SDA) and serial

A. Yu · T. Jiang (✉) · C. Zhang (✉) · Z. Yang · S. Jiang
School of Physics and Electronic Information, Yunnan Normal University, Kunming, China
e-mail: jiang_jt@126.com

C. Zhang
e-mail: zhchynnu@foxmail.com

L. C. Jain et al. (eds.), *3D Imaging Technologies—Multidimensional Signal Processing and Deep Learning*, Smart Innovation, Systems and Technologies 236,
https://doi.org/10.1007/978-981-16-3180-1_36

clock line (SCL) are needed for communication. Multiple devices can be mounted on the bus at the same time and communicate by addressing. Duplex synchronous data transmission can be realized by simple circuit connection, which saves a large number of data buses, address buses, and control buses compared with other data transmission methods. At present, the research on IIC bus mainly focuses on FPGA design, temperature monitoring, stepping motor, three-phase power, etc., while the research on IIC communication for each module in the system is less, and the test research on IIC communication protocol for embedded applications of modules is insufficient. In view of the above problems, this paper studies the IIC bus test system, proposes an IIC bus test system, and realizes the automatic switching between master and slave working modes of IIC bus communication. The IIC communication protocol can be tested quickly in the early and later verification stages of module embedded application development.

2   Circuit Hardware Design

2.1   Hardware Principle

Miller hyperbolic function model: a mathematical model of hysteresis loop suitable [3]. NJW1110 is a 9-input 3-output stereo audio selector. It includes three independent 9-input 1-output stereo audio selectors and adjustable gain buffers. NJW1110 has excellent audio characteristics, such as low distortion, low output noise, and low crosstalk. All internal states and variables are controlled by IIC bus interface. There is also a slave address selector that can be used to use the same serial bus line of both chips.

According to the composition analysis of the system, the system designed in this paper is mainly composed of STM32F103C8T6 main control module, NJW1110 channel selection module, power supply step-down module, and HMI intelligent serial port display module [4, 5], using IIC bus and serial port inside the core processor STM32F103C8T6 single chip microcomputer. The IIC bus is used to read and write NJW1110, so as to select and gain the input audio signals and control the output audio signals. The control panel adopts HMI display screen, and HMI serial port screen and STM32F103C8T6 single chip microcomputer perform read and write operation through serial port, because the power supply end of the system uses DC10V for power supply, but the power supply of STM STM32F103 single chip microcomputer is 3V3. Therefore, the circuit first uses MP2567 to step down DC10V to DC5V (because HMI display is powered by DC5V), and then uses AMS117 to step down DC5V to 3V3. The general circuit framework is shown in Fig. 1.

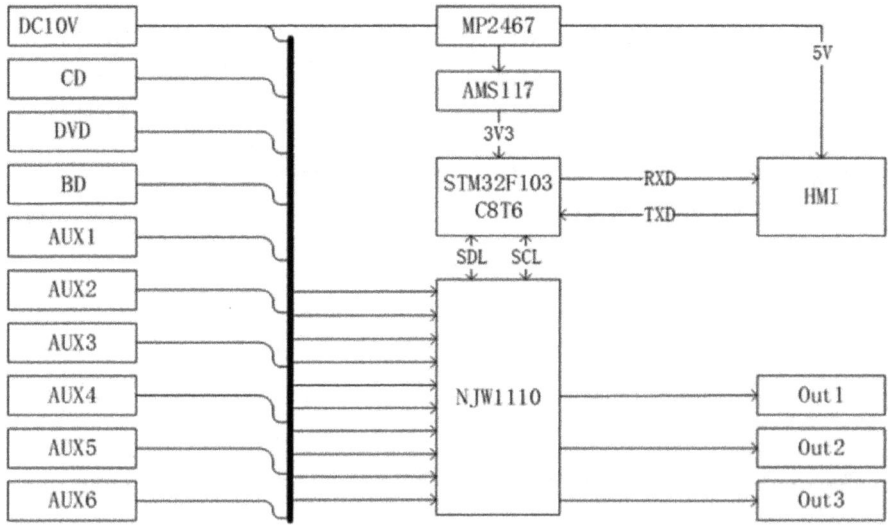

Fig. 1 Hardware frame diagram

2.2 NJW1110 Audio Selector

NJW1110 is a 9-input 3-output stereo audio selector, which includes three independent 9-input 1-output audio selectors and adjustable gain amplifiers. Its working voltage is between 7.5 and 15 V, and its working current is generally about 8 mA, with low distortion, less than 0.0007%, and extremely low output noise and crosstalk. It has built-in adjustable gain amplifier, and its gain range is adjustable from 0 to 8 dB, and its step is 0.5 dB. The main control chip accesses the register of NJW110 through IIC bus, and can use STM32CubeMX hardware IIC or analog IIC to communicate. In this paper, the input and output of the audio selection circuit are all 10Uf capacitors Line coupling filtering, impedance matching for input and output, because most power amplifiers use DC10V to supply power, so the power is supplied in the same way. When using IIC bus to read and write devices, it is necessary to know the device address. By querying the selector chip manual, the pin 18 of the chip is pulled high, and the device address when reading and writing the chip is 94H; otherwise, it is pulled high, and the device address is 96H. The hardware design of this system pulls the pin 18 low. Therefore, the circuit design of NJW1110 is shown in Fig. 2.

2.3 Analysis of Level Matching Circuit

The design of level matching circuit is an important part of hardware circuit design. By using level matching circuit, the problem of inconsistent power supply caused by using different operating voltage chips in hardware design can be well solved. In this

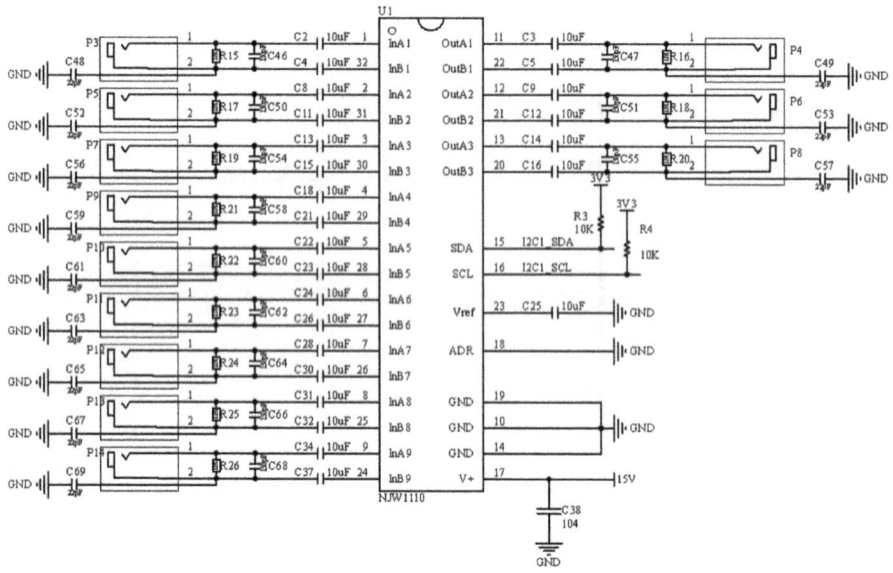

Fig. 2 Njw1110 control circuit diagram

circuit design, because NJW1110 uses DC10V to supply power, the main control chip STM32F103C8T6 (working voltage 3V3) and HMI serial port screen (working voltage 5 V) need to supply power, so the level matching circuit first uses MP2467 to reduce DC15V to 5 V, and then uses AMS117 to reduce 5 V to 3V3.

MP2467 is a high-frequency step-down integrated switching regulator, which is internally integrated with high-voltage side power MOSFET, and can provide 2.5 A output of current mode control for quick loop response and simple compensation. The input voltage ranges from 6 to 36 V, which can adapt to most step-down circuit designs. The working static current of 100 μA can be applied to battery power supply, the switching frequency is fixed at 500 kHz, and the internal soft start is achieved. The input-output conversion efficiency is as high as 95%. The output voltage can be adjusted from 0.8 to 30 V continuously. The output voltage is set by a resistor divider from the output voltage to FB pin. The divider divides the output voltage by the feedback voltage:

$$V_{FB} = V_{OUT} \frac{R2}{R1 + R2}$$

When the MP2467 is unloaded, a few microamperes of current from the high-side BS circuit can be seen at the output terminal. To absorb a small amount of current, R2 is kept at 40 k. The typical value of R2 can be 40.2 k with this, R1 can be determined as follows:

$$R1 = 50.25 * (V_{OUT} - 0.8)(k\Omega)$$

In this paper, when the output voltage is 5 V, the resistance value of R1 is 210 k, so the step-down circuit of MP2467 is shown in Fig. 3.

Ams117 5V-3V3 low dropout linear regulator is used for 5v to 3v3, which has perfect overheat protection and overcurrent protection functions. The step-down circuit diagram is shown in Fig. 4.

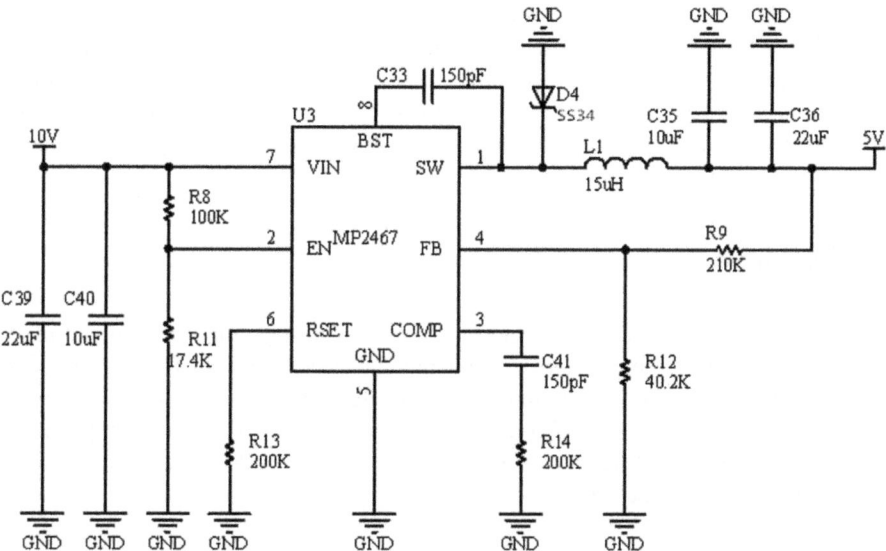

Fig. 3 MP2467 step-down circuit

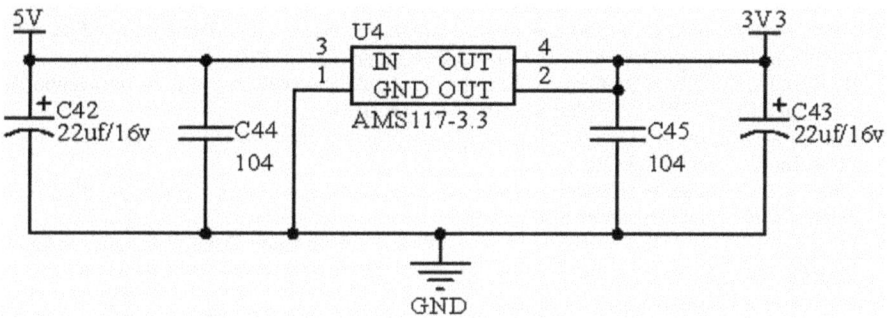

Fig. 4 AMS117 step-down circuit

3   System Test

Adjust the power supply voltage to DC10V and connect it to the system power supply, the power supply is normal and the indicator light starts flashing, use the connection cable to connect the phone audio to the DC input, select the input as DC on the serial screen, slide the volume progress bar to 100, select the output as OUT1, connect OUT1 to the audio, there is sound coming out, the sound does not show obvious distortion, adjust the volume down to 50, the sound is smaller than before, adjust the gain successfully, replace the input and output, the audio signal output is normal. Adjust the volume down to 50, the sound becomes smaller than before, adjust the gain successfully, replace the input and output, the audio signal output is normal.

4   Conclusion

This paper completes the design and production of a multi-input hi-fi preamplifier based on STM32 in terms of hardware design and system testing. Through the hardware design of IIC communication as well as algorithm design, we have gained a deeper understanding of the IIC communication method and become more proficient in the design and beautification of the HMI console using HMI as the console for human-computer interaction.

References

1. Vecchi, D., Morandi, C.: A 750 mW class G ADSL line driver with offset-controlled amplifier hand-over. In: Symposium on Mixed-Signal Design, pp. 253–258. IEEE (2003)
2. Hongpo, H., Zongfeng, Q., Jianguo, Z., Dabo, Y.: Application research of I~2C bus technology based on STM32F107VCT6 platform (05), 76–78 (2013)
3. Lan, Q., Xu, M., Jin, W.: Design of high frequency program-controlled filter. Appl. Mech. Mater. **226–228**(12), 2480–2483 (2012)
4. Chaofan, H., Yanxue, W.: Research on multi-channel signal denoising method for multiple faults diagnosis of rolling element bearings based on tensor factorization 55(12), 50–57 (2019)
5. Stott, L.N., Wicker, D.J., Marker, J.D., Rekow, A.D., Bennington-Davis, T.E., Alexander, T.: Systems and methods for using protocol information to trigger waveform analysis (2016)

Review of Constructing the Early Warning and Diagnosis Information Database of Power Plant Network Security Events

Yong Yang, Jilin Wang, Rong Li, and Jinxiong Zhao

Abstract With the rapid development of my country's economy, the country's demand for electricity is also rapidly increasing. Once the power plant network and information system have security problems, the consequences will be very serious. Therefore, studying how to power plant network and information system security has become a very important issue. Based on this, this article analyzes the development status of power plant network and information system security, summarizes the research progress of relevant industrial control system network security systems, standards, and main protection technologies at home and abroad, and introduces the composition of power plant networks and information systems. The measures related to network and information system security are analyzed and explained, aiming to ensure the security of network and information system and improve the overall operating efficiency of power plants.

1 Introduction

Electricity issues directly affect the lives of all people, and electricity safety issues also have an important impact on social stability and harmony. In recent years, with the rapid development of network information technology, power plant network information systems have gradually developed, and the development of network information tends to be integrated. People's attention must be paid to the safety of power plant networks and information systems. Once a safety problem occurs, the consequences will be very serious. Therefore, the safety protection of power plant networks and information systems has become an important task for power plants.

Y. Yang (✉) · J. Zhao
State Grid Gansu Electric Power Research Institute, Lanzhou 730070, China

J. Wang
Lanzhou Jiaotong University, Lanzhou 730070, China

R. Li
School of Science, Xi'an Polytechnic University, Xi'an 710048, China

© The Author(s), under exclusive license to Springer Nature Singapore Pte Ltd. 2021   295
L. C. Jain et al. (eds.), *3D Imaging Technologies—Multidimensional Signal Processing and Deep Learning*, Smart Innovation, Systems and Technologies 236,
https://doi.org/10.1007/978-981-16-3180-1_37

2 Security Status of Power System Network and Information System

The main contents of this part are: (1) The identification of unauthorized access channels implemented by the attacker (such as crossing the production control area and the management information area, or crossing the control area and non-control area, etc.); (2) Operation and maintenance personnel bypassing the information management area and privately accessing the production control area to implement remote operation and maintenance channels; (3) The attacker bypassing the wireless covert channel established by the firewall by setting up a wireless access point privately.

2.1 Information Security Situation in the Power Sector

At present, network and information system security is constructed by a variety of systems with different functions, all of which play an important role in controlling and managing modern power. Not only can the manpower spent on traditional power control be reduced, but the accuracy of control management is also significantly improved. However, while it brings convenience to the operation of the power system, it also exposes some potential safety hazards. Therefore, it is very important to consider the security issues existing in modern networks and information systems, which seriously affect the normal operation of the power system, and also have an important impact on the safety and stability of the entire power plant operation [1]. In recent years, the frequent occurrence of international power system network security incidents has further confirmed this trend. Electricity, as an important infrastructure, has been designated by many countries as the preferred target of "cyber warfare". The network security situation of the power monitoring system is extremely severe, and the security supervision of the cyberspace needs to be strengthened [2].

With the frequent occurrence of dangerous accidents in the power systems of countries around the world, Internet information security issues have also been exposed. Statistics show that industrial systems are currently the most attacked systems. However, information systems and networks in key areas are subjected to tens of thousands of malicious heuristic attacks every day. Such high-frequency network attacks make the information security situation particularly severe.

2.2 Research Status of Network Security Protection Technology

In view of the importance of industrial control system network security, a lot of research work has been carried out in its architecture design and protection technology at home and abroad, such as the introduction of laws and regulations, the formulation of related standards, and the development of protection technologies and products.

The United States has promulgated various comprehensive legislation related to the cybersecurity of industrial control systems, and established the Industrial Control Security Emergency Team (ICS-CERT). Six key security laboratories including the Idaho National Laboratory have been established, forming a relatively complete industrial control system network security management and technical system. The "Industrial Control System Security Guidelines" (NIST SP800-82) formulated by the NIST clarifies the typical system topology of industrial control systems. Typical threats and vulnerabilities are pointed out by this standard, and at the same time recommended security countermeasures are given [3]. Europe has formulated a cyber information security strategy by strengthening cooperation among member states in cyber security. Germany's Siemens and Schneider Electric provided technical consultation, related safety products, and a complete set of solutions [4]. The ISA and the IEC issued the "Industrial Process Measurement, Control and Automation Network and System Information Security" (ISA/IEC62443) standard specifically for industrial control network security, covering all relevant stakeholders to achieve comprehensive security protection of industrial control system network [5].

The cyber security of industrial control systems has also received extensive attention in China. In 2016, the "Cyber Security Law of the People's Republic of China" was promulgated, which clearly stated that the operation of critical information infrastructure must be secured. In terms of standard formulation, in recent years, nearly 30 related standards have been formulated for industrial control system network security, which basically covers the standard requirements of industry supervision, industrial control and safety protection product supply, industrial control owners, and evaluation agencies [6]. In view of the characteristics of the electric power industry, relevant national departments have determined the overall framework of the safety protection system of the electric power monitoring system, refined the overall principles, and provided a guarantee basis for the network security of the industrial control system in the electric power industry [7].

In recent years, the research of industrial control system network security protection technology and the development of related products have mainly focused on defense in-depth, including industrial firewalls, intrusion detection, vulnerability scanning and mining, security auditing, and situational awareness.

3   Composition of Power Plant Network and Information System

The composition of the power plant network information structure mainly includes the following parts. According to the system level correlation, it can be divided into: power plant process automatic control system (DCS), power plant information monitoring system (SIS), power plant information management system (MIS), power plant video monitoring system, power plant access control monitoring system and power plant information decision system (DSS) [8].

3.1 Power Plant Process Automatic Control System (DCS)

DCS is responsible for the collection and control of relevant data in the production process of power generation, including the main production equipment such as unit units, conveying coal powder, dust removal, and related auxiliary equipment. The system provides data about the operation of each device and is the basis for the operation of other systems.

3.2 Power Plant Information Monitoring System (SIS)

SIS is responsible for the collection and processing of monitored data, including monitoring and information query of the operating status of each device, analyzing the operating performance and status of the device, and optimizing the distribution of related loads.

3.3 Power Plant Information Management System (MIS)

MIS has two functions: one function is to optimize the balance during the construction period of the system, mainly focusing on the safety quality and construction schedule of the construction project. Another role is to focus on the safe, stable operation and economic management of the system during the operation period of the system.

3.4 Power Plant Information Decision System (DSS)

DSS is an auxiliary decision-making tool for safety operation analysis, cost economic analysis, and auxiliary analysis of power generation bidding. The main contribution is to optimize the cost of power generation while ensuring the safety of power generation.

4 Major Risks Faced by Power Plant Networks

4.1 Main Security Risks

1. Physical risk. Physical risks mainly refer to various security risks faced by information network hardware equipment, such as lightning strikes, fires, electromagnetic interference, man-made damage, etc., which will directly cause equipment damage and cause irreparable losses. In power plants, it is necessary to strengthen the security protection of the environment in which the computer network is located, and minimize the appearance of physical risks.
2. Cyber risk. Cyber risk refers to the threats from viruses and hackers faced by information networks, especially open networks, which are more vulnerable to external attacks and intrusions. Malicious attacks are carried out on the power plant information network, which leads to the paralysis of the network, and the tampering or theft of confidential information.
3. System risk. There are various systems in the power plant information network, such as operating systems, application systems, server systems, etc., and in these systems, there are bound to be certain loopholes, forming huge security risks. Once used, that will Bring immeasurable losses to the power plant [9].

4.2 Types of Attacks

The main attacks faced by power plants are: traditional network attacks, invisible penetration and scanning, hidden camouflage communications, and application layer vulnerability attacks. Among them, the latter types of attacks are the most difficult to find, so they are also the easiest attack methods used by hackers.

5 Construction of Network Security Incident Warning and Diagnosis Information Database

5.1 Build a Robust Network Architecture

1. When designing the control system, it should be in accordance with the national standard "Basic Requirements for Information Security Technology Network Security Level Protection" (GB/T22239-2019). The control system with a security level of 3.0 should be designed in accordance with the requirements of the network environment, communication network, regional boundary, computing environment, construction management, etc. [10].
2. The protection of the boundaries of different security areas is optimized in accordance with the requirements of the structural safety framework of the

power monitoring system, and the protection equipment is deployed according to the optimized structure [11]. Industrial firewalls are deployed between the controlled area (safe area I) and non-controlled area (safe area II) of the production area.

3. Generally, the principle of minimum system configuration is adopted to turn off unnecessary USB, CD-ROM, network and wireless interfaces in the control system and local smart devices, and avoid selecting the default state [12].

5.2 *Adopt Proactive Defensive Measures*

1. One of measures is to minimize the installation of the host computer (Operator station, engineer station, history station, etc.), operating system, and application software to reduce unnecessary applications, using industrial control host protection software to strengthen the security of its underlying system and improve the baseline level; The staff should carry out effective access control to files, processes, registry, and services, the purpose of which is to record and audit all system activities and user activities [13]. The management and control software adopts a process whitelist mechanism to prevent all applications and processes outside the whitelist from executing. A series of measures, including: comprehensively monitoring the host's process, network port and USB port status, deleting and deactivating default accounts and redundant and expired accounts, are all to avoid the existence of shared accounts [14].

2. When the equipment's control system is transformed and upgraded, vulnerability scanning, mining, risk assessment, and safety testing are carried out. In this way, potential security vulnerabilities, hidden dangers, and risks can be discovered in time and rectification can be carried out in time.

3. Generally, security protection equipment and systems such as intrusion detection, security auditing, and situational awareness are selectively deployed based on the system risk assessment status, industrial control system structure, and communication protocol characteristics. The mainstream approach is to access in bypass mode, and actively collect network data of industrial control systems for analysis through data flow mirroring technology. In this way, network attacks and abnormal spread of viruses can be detected in time, all suspicious data packets can be intercepted, and the system can be protected from attacks [6].

6 Conclusion

The security assurance of power plant networks and information systems is a very complex task. There are many factors to be considered, including multiple aspects and multiple levels. The guarantee of network security for power plant networks and information systems is not static, and there is a constant process of change.

Therefore, it is a very important task to construct a network security event warning and diagnosis information database for power plants. With the security event warning and diagnosis information database, existing attacks can be discovered based on rules. In this article, we mainly introduce the two important processes of network security time information database construction, how to build a robust network architecture and what active defense measures to adopt.

References

1. Lopes, J.P., Hatziargyriou, N., Mutale, J., Djapic, P., Jenkins, N.: Integrating distributed generation into electric power systems: a review of drivers, challenges and opportunities. Electr Power Syst. Res. **77**(9), 1189–1203 (2007)
2. Greenberg, A.: What Stuxnet's exposure as an American weapon means for cyberwar. Forbes.com 40–40 (2012)
3. Knowles, W., Prince, D., Hutchison, D., Disso, J.F.P., Jones, K.: A survey of cyber security management in industrial control systems. Int. J. Crit. Infrastruct. Prot. **9**, 52–80 (2015)
4. Skare, P.M.: Method and system for cyber security management of industrial control systems. US Patent. 8595831 (2013)
5. Robinson, C.: Standards update: IACS cybersecurity. In Tech **61**(5) (2014)
6. An, G.F., Zhu, C.M., Lei, X.F., Li, Y.N.: Information security policy and standard system of industrial control system in China. Laws Regulations **4**(10), 959–964 (2018)
7. Huan, Y., Songhua, L., Lifang, H.A.N.: Overview of power industry control system security technology. Electr. Power Inf. Commun. Technol. **16**(03), 56–63 (2018)
8. Luo, F.Z., Wei, W., Zhang, Y., Li, F., Liu, Z.C., Xu, Y.H., Qu, J.H., Wang, Y.: Study and application of integrated power network planning information system. In: 2012 Asia-Pacific Power and Energy Engineering Conference, pp. 1–5. IEEE (2012)
9. Yaacoub, J.P.A., Salman, O., Noura, H.N., Kaaniche, N., Chehab, A., Malli, M.: Cyber-physical systems security: limitations, issues and future trends. Microprocess. Microsyst. **77**, 103201 (2020)
10. Chen, G.Y., Zhu, G.B., Fan, C.L.: Information security technology—evaluation requirement for classified protection of cybersecurity (GB/T 28448–2019) standard interpretation. Netinto Secur. **19**(07), 1–7 (2019)
11. Yang, H., Luo, H., Ye, F., Lu, S., Zhang, L.: Security in mobile ad hoc networks: challenges and solutions. IEEE Wirel. Commun. **11**(1), 38–47 (2004)
12. Li, J.Q., Yu, F.R., Deng, G., Luo, C., Ming, Z., Yan, Q.: Industrial internet: a survey on the enabling technologies, applications, and challenges. IEEE Commun. Surv. Tutorials **19**(3), 1504–1526 (2017)
13. Knapp, E., Broad, J.: Industrial network security: securing critical infrastructure networks for Smart Grid, SCADA, and other industrial control systems. Syngress. (2014)
14. Zalewski, J.: The practice of network security monitoring: understanding incident detection and response. Comput. Rev. **55**(5), 260–261 (2014)

Improvement and Application of the Prediction Algorithm of Atomic Clock Combination Clock Difference

Yingying Gong, Gaochang Zhao, and Tian Hui

Abstract The clock error prediction of satellite navigation system requires high precision and strong prediction stability. A combined clock error prediction model is designed in this paper. The background value weight is set to the dynamic optimization variable, and the initial value is improved to optimize the grey model (GM(1,1)). At the same time, the network weight and thresholds of BP neural network are determined by particle swarm optimization algorithm to optimize the prediction performance. Then, the optimized preliminary prediction results of different GM(1,1) are taken as training values and input into the optimized BP network to obtain the final prediction results. The simulation experiment is carried out by using International GNSS Service organization precise clock difference data and compared with grey forecasting model before and after improvement, the quadratic polynomial model, the BP neural network model, the grey neural network model and the forecasting model in this paper, respectively. The results show the advantages of the combined prediction model and meet the requirements of real-time and high-precision clock error prediction of satellite navigation system.

1 Introduction

In the satellite navigation systems, the atomic clocks on board is very sensitive because of its high frequency, which is easily influenced by the outside world and its own factors. The results show that the grey model [1] has the advantages of flexible modelling and high prediction accuracy; it is suitable for short-term prediction. Still, its prediction accuracy requires that the clock error data be exponential and has limitations. BP neural network [2] has the characteristics of fast learning speed and superior generalization performance, which is suitable for nonlinear time series prediction. However, it sometimes does not perform a good global search and fast convergence. The grey systems and neural networks [3] are parallel in the literature [3]. The model is more dependent on sample data, and too little sample data may be

Y. Gong (✉) · G. Zhao · T. Hui
College of Science, Xi'an University of Science and Technology, Xi'an, China

L. C. Jain et al. (eds.), *3D Imaging Technologies—Multidimensional Signal Processing and Deep Learning*, Smart Innovation, Systems and Technologies 236,
https://doi.org/10.1007/978-981-16-3180-1_38

very limited. In Ref. [4], the grey system and the neural network are connected in series, and the forecast values obtained by using different grey models are taken as the final output forecast values of the sample set of the neural network [4]. In literatures [5, 7] the grey system is "embedded" in the neural network [5, 7], and the weight and threshold parameters of the neural network are optimized by the intelligent optimization algorithm. The model solves the randomness of parameter selection of grey neural network and quickly falls into the local optimum, and improves the forecast precision.

In this paper, an improved combined prediction model is designed, according to the advantages of series grey neural network and the defects of error accumulation. It is mainly improved from four aspects: the improvement of background value construction, the reconstruction of whitening equation parameters, the modelling of the grey differential equation and the direct calculation of parameters. This paper sets the weight of the background value as a variable for dynamic optimization and improves the initial value [8]. The improved GM(1,1) clock difference prediction value is trained and the original clock difference data is input into BP network [9] as the test value. The final prediction results are obtained by combination prediction.

2 Algorithm Introduction

2.1 Optimized GM(1,1) Model

Theoretical research shows that the average background value of the traditional GM(1,1) model is defective, and the prediction accuracy is not necessarily the highest. The traditional GM(1,1) model uses the mean background value structure, but theoretical research shows that the mean background value has defects and the forecast accuracy is not necessarily the highest.

Suppose the original clock error data is

$$x^{(0)} = \left(x^{(0)}(1), x^{(0)}(2), \ldots, x^{(0)}(n)\right) \tag{1}$$

The $x^{(0)}(i), i = 1, 2, \ldots n$ is the original clock error data. The formula (1) is accumulated once to obtain the following formula

$$x^{(1)} = \left(x^{(1)}(1), x^{(1)}(2), \ldots, x^{(1)}(n)\right) \tag{2}$$

The mean background value of the first-order accumulation generated sequence is

$$z^{(1)}(k) = 0.5\left(x^{(1)}(k) + x^{(1)}(k-1)\right) \tag{3}$$

Now construct the weighted background:

$$z^{(1)}(k) = \lambda x^{(1)}(k) + (1 - \lambda)x^{(1)}(k - 1) \tag{4}$$

The λ is the background weight. Set the background value weight as a variable for dynamic optimization modelling:

$$\lambda = \lambda + \Delta\lambda, \lambda \in (0, 1), \Delta\lambda = 0.01 \tag{5}$$

We set the initial values $\hat{x}^{(1)}(1)$ and adjust the parameters C:

$$\hat{x}^{(1)}(1) = x^{(0)}(1) = c + \frac{\mu}{a} \tag{6}$$

$$C = c \cdot \left(1 - e^a\right) \tag{7}$$

where the a is the development coefficient, the μ is the action quantity and the c is initial conditions. Substitute the grey differential equation

$$\hat{x}^{(1)}(k + 1) = C \cdot \left(1 - e^a\right)^{-1} \cdot e^{-ak} + \frac{\mu}{a}, k = 0, 1, \ldots, n - 1 \tag{8}$$

$$\hat{x}^{(0)}(k + 1) = \hat{x}^{(1)}(k + 1) - \hat{x}^{(1)}(k) = C \cdot e^{-ak}, k = 1, 2, \ldots, n - 1 \tag{9}$$

$$C = \frac{\left[x^{(0)}(1) - \frac{u}{a}\right](1 - e^a)^{-1} + \sum_{k=2}^{n} x^{(0)}(k)e^{-a(k-1)}}{(1 - e^a)^{-2} + \sum_{k=2}^{n} e^{-2a(k-1)}} \tag{10}$$

According to the principle of the minimum sum of squared deviations between the predicted value and the actual value, S, the optimal background value weight and initial value are obtained:

$$S = \sum_{k=1}^{n} \left(\hat{x}^{(0)}(k) - x^{(0)}(k)\right)^2 \tag{11}$$

2.2   BP Neural Network

BP three-layer network topology of neural network model includes input layer, hidden layer and output layer. It uses the backpropagation of one layer by one of the fastest descent methods to constantly adjust the weights and biases of the network. The essence of its training and learning is the dynamic adjustment of each connection weight. BP network is actually a multilayer perceptron, so its topology is the same

Fig. 1 Three-layer network topology

Input Layer   hidden layer   output layer

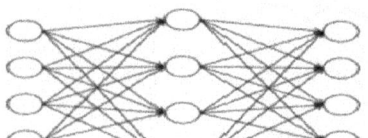

as that of multilayer perceptron. Because the three-layer perceptron can solve simple nonlinear problems, it is most widely used. BP three elements of the network are network topology, transfer function, and learning algorithm. A three-tier BP network topology is shown in Fig. 1.

BP neural network uses existing data to repeatedly test and train, find the approximate weight relationship between input and output, and use the obtained weight relationship to simulate prediction. When the output of the neural network is far from the expected output, there is an output error. The size of the error can be changed by adjusting the weight of each layer. Transfer function used in the BP neural networks is

$$f(x) = \frac{1}{1 + e^{-x}} \text{ or } f(x) = \frac{1 - e^{-x}}{1 + e^{-x}} \tag{12}$$

2.3 Optimization of BP Neural Networks by Particle Swarm Algorithm

The traditional BP neural networks have limitations and need to optimize their weights and biases. Step 1: The connection weights of each layer of the neural network are encoded into particles, the number of nodes in the input layer, the hidden layer and the output layer of the BP neural network is determined, and the connection weights and bias lengths of the BP neural network are initialized as the coding of the particle swarm. Step 2: Initialize the population size, maximum iteration times, particle position, maximum and minimum of particle velocity, and introduce inertia weight and time factor. Step 3: Through network training, calculate the fitness function value of particle. Step 4: When the objective function of the training network reaches the optimum, the iteration is stopped, and the result is obtained. Otherwise, the iteration continues until the algorithm converges. Step 5: Finally, the weights and thresholds of the BP neural network are determined, GM(1,1) different predicted values are used as training samples, and the original clock error data are used as test samples to train and test the neural network and output the predicted values.

3 Experiment and Result Analysis

The time period provided by International GNSS Service organization is selected as the precise clock bias data from 00:00 on May 24, 2016, to 23:55 on May 25, 2016. The precise clock difference data of four satellites numbered G01, G15, G20, and G32 were taken as samples, and the sampling interval was 5 min, a total of 576 groups of data. In this paper, the mean square error (RMSE) and the extreme difference (Range) are used to characterize the accuracy and stability of each prediction model:

$$\text{RMSE} = \left[\frac{1}{n} \sum_{i=1}^{n} \left(\text{error}^{(i)} \right)^2 \right]^{1/2} \tag{13}$$

$$\text{Range} = \left| \text{error}_{\max}^{(i)} - \text{error}_{\min}^{(i)} \right| \tag{14}$$

Among them $\text{error}^{(i)} = \hat{x}_i - x_i$, for each model forecast error; $\text{error}_{\max}^{(i)}$ and $\text{error}_{\min}^{(i)}$, respectively, represents the maximum error and the minimum error of each model forecast error.

This paper uses the first six hours of original clock difference data modelling to predict the clock difference of the next six hours, 12 h, 24 h, and 42 h, respectively, and compares and analyses the model with the original precision data. In this paper, an improved grey neural network model (IGM-BP) is presented. Firstly, different GM (1,1) are established by using the data of the first six hours, the second five hours, the second four hours and the third three hours. Then different GM(1,1) prediction results are input into the improved BP neural network as training values to predict the clock difference of the next six hours, 12 h, 24 h, and 42 h, respectively. Finally, it is compared with the GM(1,1) before and after improvement, quadratic polynomial model, BP neural network model and grey BP neural network model, respectively. The mean square error and range are used to evaluate the accuracy and stability of each model. Table 1 shows the mean square error results of clock errors predicted by different models for four satellites for the next six hours, 12, 24, and 42 h. Figure 2 shows the four satellite clock error prediction curves.

The GM(1,1) model before and after the improvement has better prediction accuracy for clock difference in the next six hours and twelve hours, the prediction accuracy is less than 0.6 ns, which is 30–50% higher than before the improvement. Although the improved model improves the prediction accuracy, it is still unstable. Grey neural network combined prediction model has a clock error prediction accuracy of fewer than 0.5 ns in the next six hours, and a clock error prediction error of fewer than six nanoseconds for the next 12 h and the next 24 h. With the improvement of grey BP neural network model, the prediction accuracy is increased by 20–60, the clock error prediction accuracy for the next 6 h is less than 0.3 ns, and the clock error prediction performance of the next 12, 24, and 42 h is also superior to other methods.

Table 1 Mean square error of clock error of four satellites ns

Clock	Time (h)	GM	IGM	QP	BP	GM-BP	IGM-BP
G01	6	0.1302	0.0958	0.8186	11.3366	0.3135	0.0238
	12	0.2169	0.1641	2.0771	56.2144	0.4290	0.0249
	24	0.9597	0.8512	6.0734	71.7390	0.7735	0.2590
	42	1.3450	1.2246	17.8374	11.4506	1.2323	0.6260
G15	6	0.5667	0.4213	3.4771	33.4559	1.1820	0.3016
	12	0.9488	0.8607	8.6293	20.6691	1.9747	0.3494
	24	1.8533	1.6788	25.9689	72.5314	3.8596	1.3551
	42	3.0168	2.8407	68.8625	89.1401	6.4289	1.6287
G20	6	0.2949	0.2089	1.3436	18.1030	0.4662	0.0859
	12	0.6541	0.4587	3.5158	99.7136	0.9865	0.2455
	24	1.4299	1.3775	10.5026	101.5933	2.0450	0.9785
	42	2.2176	1.9805	27.0292	131.9935	3.2274	1.3296
G32	6	0.5317	0.3954	0.4608	40.7886	0.3698	0.1325
	12	0.9612	0.9284	3.0907	73.4670	1.1990	0.4873
	24	4.8139	4.2758	12.1118	95.0485	5.4305	1.3610
	42	15.4451	14.7995	35.1851	112.8349	16.7220	14.0952

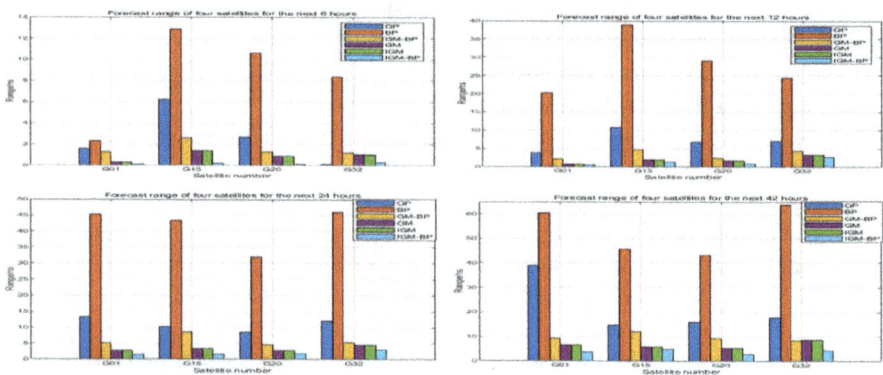

Fig. 2 Four satellite clock error prediction curves

The extreme difference in the BP neural network model is massive, and the forecast is very unstable. The quadratic polynomial model has poor prediction stability for clock errors in the next 6 and 12 h, and lower for clock error prediction in the next 24 and 42 h. Before and after the improvement, the stability of the GM(1,1) model is better, but with the increase of the forecast time, the stability of the forecast decreases gradually. The prediction stability of grey BP neural network model for clock difference in the next six hours and 12 h is similar to that of the grey model. Still, the prediction stability of clock difference in the next 24 and 42 h in general. By contrast, the improved grey BP neural network model has better prediction performance, especially in the medium- and long-term prediction, which embodies its superiority.

As can be seen from Figs. 3, 4, 5 and 6, the prediction error of BP neural network is larger, and the clock error of the next 24 and 42 h is larger, so it is best to use

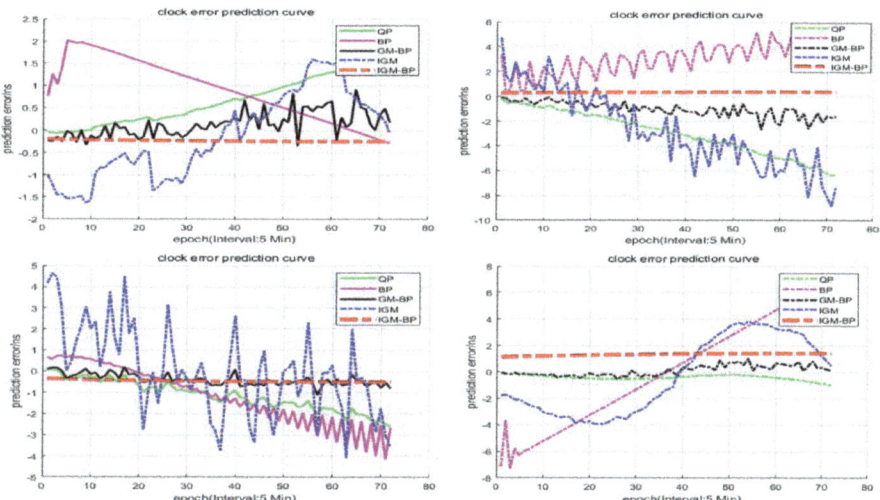

Fig. 3 Forecast error curve of four satellites in the next six hours

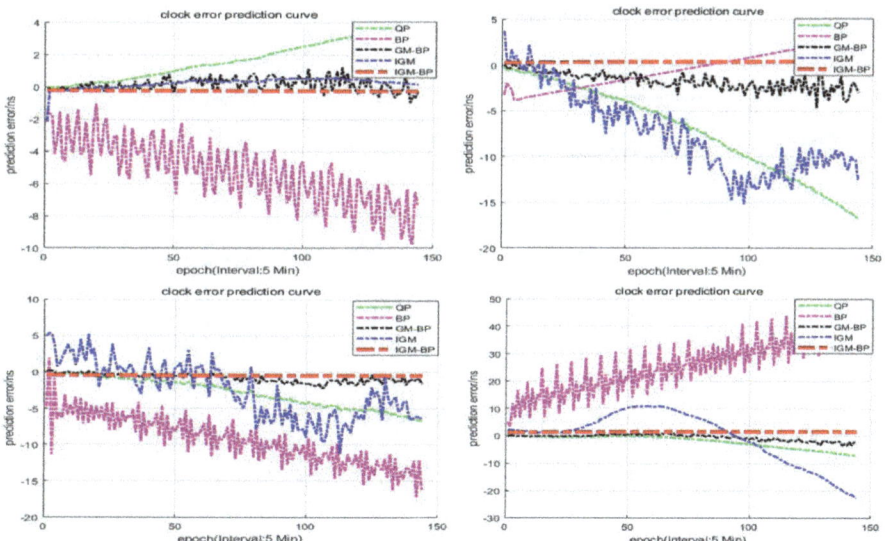

Fig. 4 Forecast error curve of four satellites in the next 12 h

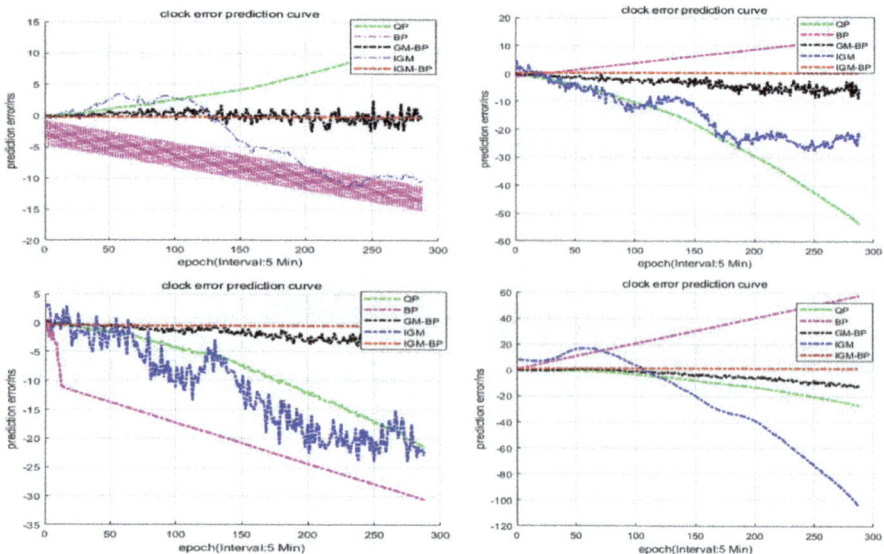

Fig. 5 Forecast error curve of four satellites in the next 24 h

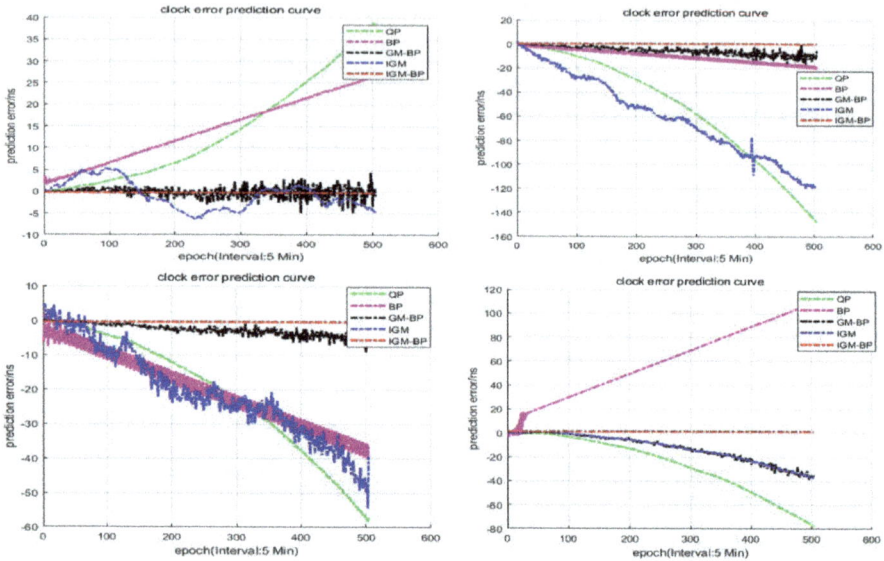

Fig. 6 Forecast error curve of four satellites in the next 42 h

the collocation optimization algorithm. The grey neural network forecasting effect is good, but there is still the phenomenon of error accumulation. The improved grey BP neural network forecasting effect is best, the forecasting error is low, and the stability is good. The error accumulation phenomenon is reduced; its advantage is also embodied in the medium- and long-term clock error prediction.

4 Conclusion

The grey neural network model has good performance for clock difference prediction in the next six hours and 12 h, and the clock difference forecast for the next 24 and 42 h is slightly better than the grey model. Compared with other models, it also has better stability and applicability in clock difference prediction for the next six hours and 12 h. All the above prediction models have error accumulation phenomenon, and the combined prediction model in this paper can reduce this error accumulation phenomenon very well. A combined prediction model in this paper improves has good prediction performance and applicability.

Acknowledgements Thanks to the editors and reviewers for their valuable comments and suggestions. The project is supported by the Natural Science Foundation of Shaanxi Province (2019JQ-346).

References

1. Li Wei, F., Cheng Pengfei, S., Mi Jin Zhong, T.: Application of the grey system model to the short-term prediction of satellite clock error. Surv. Bull. **25**(6), 32–35 (2009)
2. Zhu Lingfeng, F., Li Chao, S., Li Xiaojie, T.: Application of neural network in clock error prediction of navigation satellite. Measur. Tech. Space Probe **36**(3), 41–45 (2016)
3. Guo Zhenhua, F., Han Baomin, S., Zhao Jinsheng, T.: Satellite clock prediction based on Grey BP network. J. Shandong Univ. Technol. (Natural Science Edition) **26**(4), 39–42 (2012)
4. Lei Yu, F., Zhao Danning, S., Gao Yuping, T.: Clock error prediction based on Grey system and neural network. J. Time Freq. **36**(3), 156–163 (2013)
5. Zou Bing, F., Chen Xihong, S., Xue Lunsheng, T.: Satellite clock error prediction based on Grey neural network optimized by drosophila optimization algorithm. Sci. Surv. Mapp. **39**(9), 44–48 (2014)
6. Yang Fan, F., Xie Yangyang, S.: Satellite clock error prediction model based on Genetic Grey Neural Network. J. Navig. Positioning **5**(2), 107–134 (2017)
7. Zhao Zengpeng, F., Yang Fan, S., Zhang Ziwen, T.: Satellite clock error prediction based on Grey Neural Network optimized by particle swarm optimization. J. Navig. Positioning **6**(2), 53–56 (2018)
8. Yang Hualong, F., Liu Jinxia, S., Zheng Bin, T.: Improvement and application of Grey prediction GM (1,1) model. Math. Pract. Knowl. **41**(23) (2011)
9. Chen Ximing, F., Huang Zhangyu, S., Qin, J., Liu Renzhi, T.: PSO optimized neural network for short-term clock error prediction. Surv. Mapp. Sci. **44**(9), 7–12 (2019)

Design of Smart Humidifier Based on Voice Control

Zewang Zhang, Chenwei Feng, Huangbin Zeng, and Yelan Chen

Abstract Nowadays air pollution is serious. A humidifier can effectively improve the ambient humidity, and the negative oxygen ions released by its atomization are beneficial to people's health. However, the function of the humidifier is relatively simple at present. An STM32 is used as the core processing unit, a speech recognition module based on LD3320 and a speech synthesis module based on SYN6288 are chosen as the control mode, a more practical and convenient smart humidifier is designed in this paper. ADHT11 temperature and humidity sensor are used to collect ambient humidity, and the collected information is displayed in OLED after MCU processing. If the current air humidity is less than the set value and the water level reaches the standard, the humidifier will begin to humidify. Voice warning is issued when the water level of the humidifier is too low. Besides, the opening and stopping of humidification can be controlled by voice. The testing results show that the humidifier designed is stable with a high-speech recognition rate.

1 Introduction

With the development of society and the economy, people also put forward higher requirements for indoor environmental comfort. A good indoor environment not only helps people's health but also brings about pleasure. The use of the humidifier can effectively change the air humidity and provide people with a comfortable environment. However, the functions of the humidifier are relatively single at present. How to make the humidifier more intelligent is worth studying further [1]. Voice control is a hot topic in the smart home. Compared with traditional manual operation, voice control greatly facilitates people's work and life [2]. Therefore, the design of a smart humidifier based on voice control will improve people's quality of life and health level, and develop the humidifier industry.

Z. Zhang · C. Feng (✉) · H. Zeng · Y. Chen
School of Opto-Electronic and Communication Engineering, Xiamen University of Technology, Xiamen, China
e-mail: chevyphone@163.com

© The Author(s), under exclusive license to Springer Nature Singapore Pte Ltd. 2021
L. C. Jain et al. (eds.), *3D Imaging Technologies—Multidimensional Signal Processing and Deep Learning*, Smart Innovation, Systems and Technologies 236,
https://doi.org/10.1007/978-981-16-3180-1_39

The humidity and water level information are collected by the corresponding sensor, and a smart humidifier based on voice control is designed in this paper. After the system is started and initialized, the air humidity and water level information can be displayed not only by OLED but also by voice broadcast. If the current air humidity is less than the set value, the system starts humidifying. If the air humidity reaches the set value, the system will stop humidifying. During the humidification process, if the humidifier has insufficient water, the humidifier will stop humidifying and send a voice alarm. During the operation of the system, the humidifier can be controlled by voice, including starting or stopping humidification and setting the desired humidity value.

2 System Design

The overall design block diagram of the smart humidifier based on voice control is shown in Fig. 1. The control core of the system is an STM32 microcontroller, which is equipped with a speech recognition module, a humidity acquisition module, and a water level detection module. The humidity acquisition module based on DHT11 [3] is used to collect the current humidity. The water level detection module based on Risym water sensor can detect the water volume of the water storage device in real-time. If the water volume is insufficient, the system will timely give an alarm to prevent the humidifier from damage due to continuous humidification. The speech recognition module can receive and recognize the user's instructions, and send the corresponding instructions to the serial port through matching mode to control the opening and closing of the atomization module. The speech synthesis module enables the system to show users various voice messages. Besides, humidity information and water level information can also be displayed through OLED.

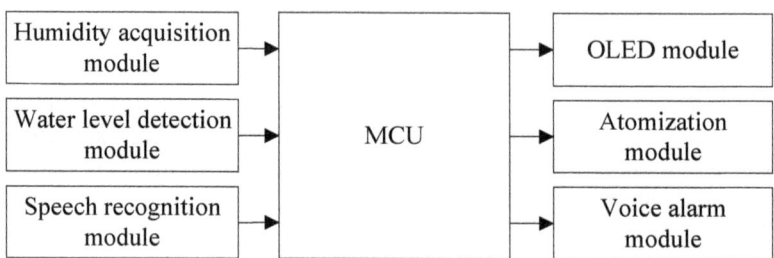

Fig. 1 System block diagram

Fig. 2 Humidity acquisition
circuit diagram

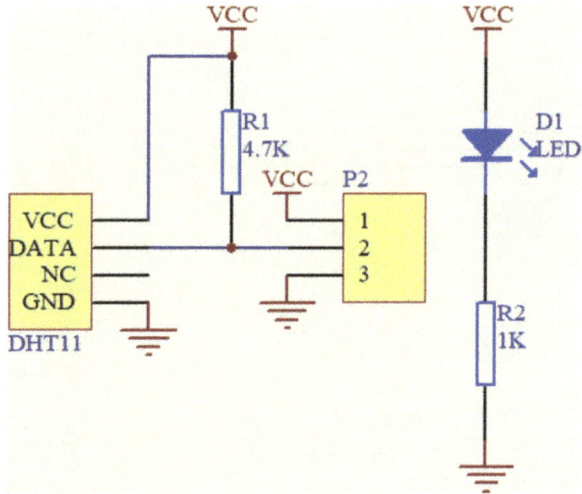

3 Hardware Design

3.1 *Humidity Acquisition Module*

The circuit of the humidity acquisition module is shown in Fig. 2. The humidity
acquisition module uses the DHT11 sensor to transfer the collected environmental
humidity information to MCU for processing. The DHT11 is a temperature and
humidity composite sensor with a calibrated digital signal output, which has high
reliability and stability.

3.2 *Water Level Detection Module*

The water level detection module based on Risym water sensor shown in Fig. 3 is
used to detect the water level of the water storage device and to warn when the water
level is below the set value, thereby improving the safety of the humidifier. The

Fig. 3 Water level detection
module

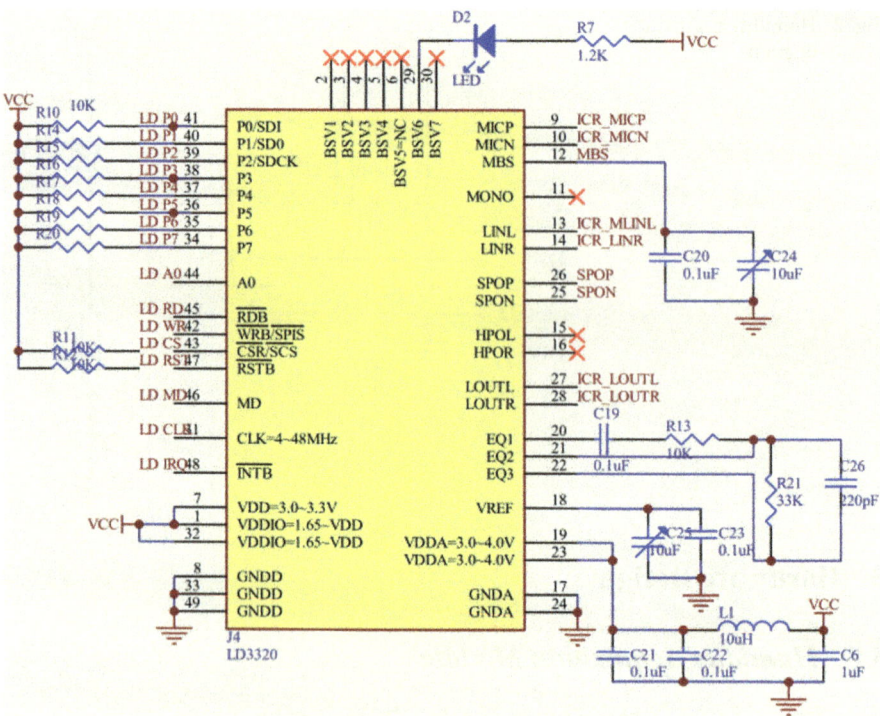

Fig. 4 Speech recognition circuit diagram

sensor works by detecting the current water level with several parallel wires exposed outside and automatically realizing A/D conversion.

3.3 Speech Recognition Module

The LD3320-based speech recognition module is used for speech recognition, the circuit of which is shown in Fig. 4. The LD3320 is a special speech recognition chip [4], which integrates speech recognition, speech processing, A/D conversion, and other functions. The result of speech recognition will be sent automatically through the serial port.

3.4 OLED Module

The OLED is used as a display module. The OLED is rich in functions and easy to operate, which has several control commands, such as brightness control, contrast control, etc. In addition, the OLED is suitable for small systems with low-power consumption, small size, and thin thickness.

3.5   Atomization Module

The humidifier uses the Eckert USB-based atomization module to atomize the water in the water storage device to improve the air humidity. The ceramic atomizer inside the atomization module sprays water mist into the air after the shock absorbed from the water storage device is dispersed by high-frequency electronic oscillation. This form of water mist also causes the water mist to carry negative ions, which can react electrostatically with large and small particles in the air, causing them to accumulate and precipitate [5].

3.6   Voice Alarm Module

The system uses a speech synthesis module based on SYN6228 for voice alarm [6], and the module circuit is shown in Fig. 5. The SYN6228 receives the text to be synthesized through an asynchronous serial UART and then converts the text to sound (TTS). The SYN6288 supports multiple text formats, multiple baud, and 16 volume level adjustments.

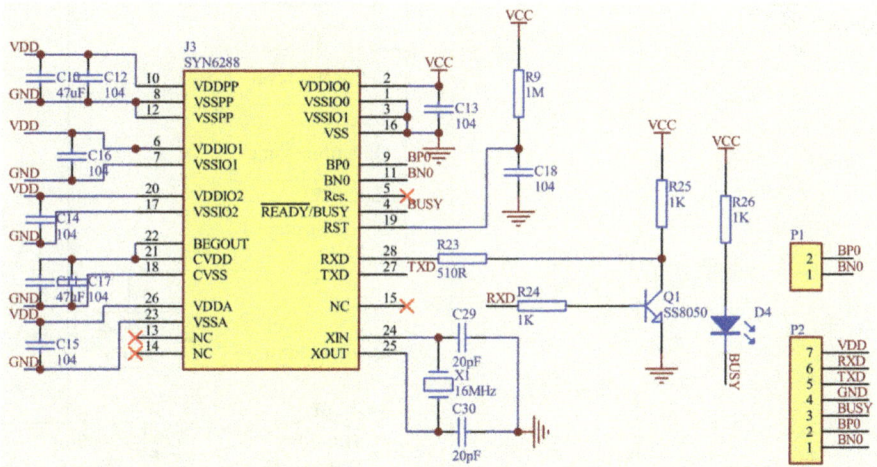

Fig. 5 Voice alarm circuit diagram

4 Software Design

The system software program includes the main program and serial port interrupt program. The main program flow of the system is shown in Fig. 6. After initialization, the system enters the cycle process. Firstly, the sensor is started to read the current environmental humidity and water level information of the water storage equipment, and the data is processed and sent to the OLED display. Then, judge whether the water level is greater than the lower limit. If the water level is too low, the system starts the voice alarm module to prompt the user to add water. If the water level is greater than the set value, judge whether the current humidity is less than the set value. If the environmental condition is too dry, turn on the atomizer for humidification; if the current ambient humidity is greater than the set value, turn off the atomizer.

The system enters the interrupt procedure flow as shown in Fig. 7 through serial port interrupts and operates by using speech recognition results, including functions of starting humidification, stopping humidification, and setting humidity. Every time the voice control is carried out, the speech recognition module needs to wake the system up through the first level command "bottle." After the humidifier gives the

Fig. 6 Main program flow chart

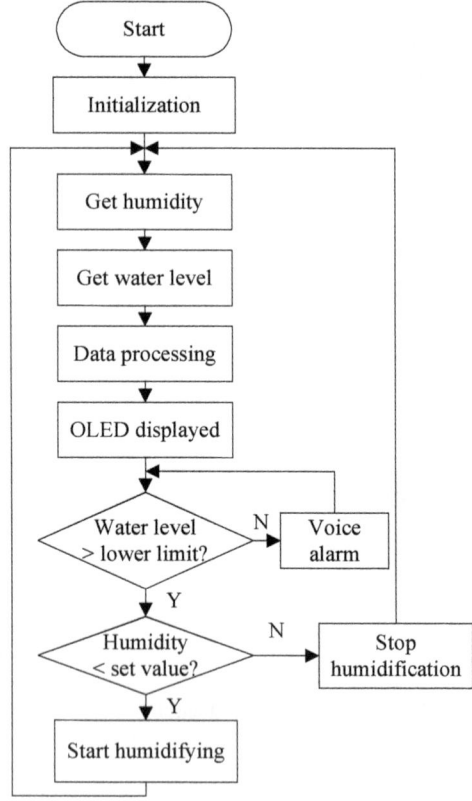

Fig. 7 Interrupt program flow chart

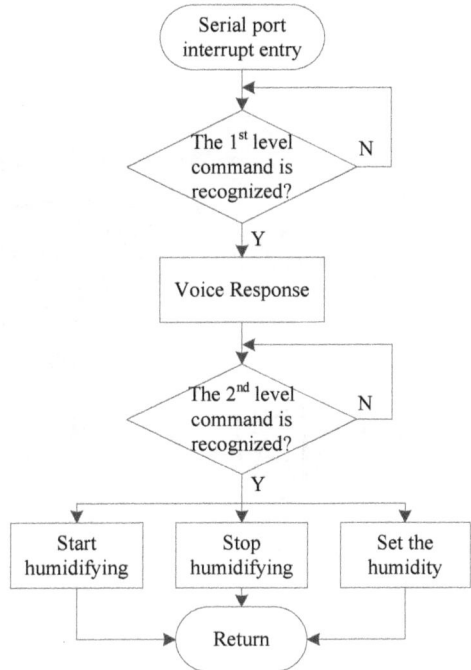

voice response "I am," it can be controlled by various voice commands, such as "start humidification," "stop humidification," "set humidity to 30," etc. The MCU executes the corresponding operation by judging the instruction and returns to the main program after the interrupt program ends.

5 Test and Analysis

The system photo shown is shown in Fig. 8. In order to test the performance of the system proposed, a Xiaomi system based on Xiaomi smart speaker and Xiaomi humidifier on the market is selected for comparison [7].

Firstly, the accuracy of speech recognition for these two systems in quiet and noisy environments is tested. When the environment is quiet, the accuracy of speech recognition for the system proposed can reach to 95%. When the environment is noisy, the accuracy is 83.3%. The accuracy of speech recognition for Xiaomi system in different environments is lower than that of the system proposed, respectively, which is 93.3 and 81.6%. Then, the response time is tested in quiet environment. The testing results show that response time for the system proposed is about 2 s and the Xiaomi system is about 3 s. The reason is that the system proposed with smaller delay does not need to access Wi-Fi.

Fig. 8 System photo shown

6 Conclusion

To enhance functions of the humidifier and reduce the delay of the response time for voice control, a smart humidifier based on offline voice control is designed in this paper. The humidifier can start humidifying when the ambient humidity is lower than the set value and can stop humidifying when the humidity is reached to the set value. In order to prevent the humidifier from working when the water level is too low, this design has set up the anti-dry burning alarm module. When the water level is lower than the set value, the humidifier will stop working and give an alarm through the speech module. In addition, the humidifier can also be controlled by voice commands, and the delay is smaller than the online system in the market. The testing results show that the system works stably with high-speech recognition rate and low delay. The proposed system provides people with a more convenient and intelligent interactive experience to improve the quality of life.

Acknowledgements This work was supported by Education and Scientific Research of Young Teacher of Fujian province (Grant No. JT180458, No. JT180430, and No. JAT190677), Scientific Research Climbing Project of Xiamen University of Technology (Grant No. XPDKT19006), and High-level Talent Project of Xiamen University of Technology (Grant No. YKJ17021R).

References

1. Ohta, Y., Kasamatsu, K.: Characteristics of the appearance and uniqueness of humidifiers. In: 2013 International Conference on Biometrics and Kansei Engineering, pp. 125–128. Tokyo (2013)
2. Nan, E., Radosavac, U., Papp, I., Antić. M.: Architecture of voice control module for smart home automation cloud. In: 2017 IEEE 7th International Conference on Consumer Electronics—Berlin (ICCE-Berlin), pp. 97–98. IEEE, Berlin (2017)
3. Wang, L.: Design of speech recognition system based on LD3320 chip. In: 2016 3rd International Conference on Materials Engineering, Manufacturing Technology and Control. Atlantis Press (2016)
4. Wang, P., Lu, X., Sun, H., Lv, W.: Application of speech recognition technology in IoT smart home. In: 2019 IEEE 3rd Advanced Information Management, Communicates, Electronic and Automation Control Conference (IMCEC), pp., 1264–1267. IEEE, Chongqing, China (2019)
5. Halim, A.H.H., Ohyama, R., Tsuji, T., Yoshida, T.: Influence of electrostatic induction electrode configuration on water mist charging. In: 2011 Annual Report Conference on Electrical Insulation and Dielectric Phenomena, pp. 231–234. Cancun (2011)
6. Shunxia, C., Yanda, C.: Design of wireless intelligent home alarm system. In: 2012 International Conference on Industrial Control and Electronics Engineering, pp. 1511–1513. Xi'an (2012)
7. Feng, C.W., Xie, H.M.: The smart home system based on voice control. In: 2019 Annual International Conference on 3D Imaging Technology, pp. 383–392. IC3DIT, Kunming, China (2019)

Image Target Detection Method Using the Yolov5 Algorithm

Shengxi Jiao, Tai Miao, and Haitao Guo

Abstract Human fall detection is a problem to be solved in video processing, and it is applied to the intelligent transportation and the smart medical care. A new method that combines the yolov5 algorithm and the ResNet-50 network is proposed in this paper to realize real-time human fall detection in video images. First, the yolov5 algorithm is used to detect moving targets in video images, and then the detected images with moving human targets are preprocessed. Furthermore, the processed images are sent to the convolutional neural network-ResNet 50 to perform the classification of determining whether the human body falls. In this paper, ResNet-50 network uses UR Fall Datasets—an online open dataset—to process video frames into images, as sample data for training. Results show that the accuracy of the algorithm proposed in this paper can reach 93.9%, which has a good application effect in the field of fall detection.

1 Introduction

Human fall detection belongs to the category of video processing, and it is applied to the intelligent transportation and the smart medical care. At present, human fall detection is mainly divided into three categories: wearable devices, environmental sensors, and computer vision-based methods. For wearable devices, gyroscopes, three-axis accelerometers, and other devices are mainly used to measure human movement indexes. Mathieet et al. [1] uses the waist wearers to extract movement parameters with accelerometers, and then analyzes whether a human falls. Bianchi et al. [2] adopted atmospheric pressure sensor to collect pressure data, together with motion data collected by accelerometer, and adopted heuristic training decision tree classifier to carry out fall detection. Tamura et al. [3] designed an airbag that will

S. Jiao (✉) · T. Miao
Northeast Electric Power University, Jilin 132012, China

H. Guo
Hainan Tropical Ocean University, Sanya 572022, China

be triggered when the acceleration and angular acceleration detected by the wearing device are greater than a given threshold, which is of great significance in protecting the human body from falls. Although the method based on wearable sensors has high accuracy [4], it is difficult for the elderly to popularize wearable devices because they are not only highly repellent, but also always forget the reason for wearing them. And environmental sensors not only home configuration are more difficult, but also easy to cause errors caused by detection; detection accuracy is not enough and unsuitable for promotion at present. Therefore, the method of computer vision can solve this problem very well. Chua et al. [5] proposed to divide the human body into three parts, namely the head, trunk, and feet. Agrawal et al. [6] proposed an improved GMM as background subtraction method to identify whether the human template was a human object based on contour matching, and to detect falls by calculating the aspect ratio of the human body and the midpoint distance. Liu et al. [7] proposed a fast three-frame background modeling method to detect falls through aspect ratio of human body, centroid velocity, head movement distance, and final posture Angle. Yao et al. [8] proposed to use two fitting ellipses to fit the head and trunk of the human body respectively, and finally use shallow CNN to analyze the correlation between the contours of the two ellipses according to the features extracted from the ellipses to perform fall detection. Rougier et al. [9] proposed the fall detection based on the combination of motion history (MHI) and human shape change.

At present, the accuracy of the fall detection using video is low. This paper proposes a method for fall detection using yolov5 combined with the ResNet-50 [10].

2　Methodology

In this study, in order to study yolov5 combined with ResNet-50 for moving target detection method and apply it to the human body fall detection field. Firstly, in this paper, yolov5 algorithm is used to detect moving human objects in video images, and then to detect human movement, target image is preprocessed. Finally, the processed images are fed into the ResNet-50 network for classification to determine whether the human fall or not.

2.1　Development Environment for System

The algorithm proposed in this paper uses the NVIDIA GeForce GTX 1660 TI graphics card acceleration, 8 GB memory, Intel(R) Core(TM) i5-9300H CPU @2.40 GHz laptop running code, which is calculated on the Ubuntu 18.04 operating system, Python version is 3.6.6, opencv-python = 3.4.2, PyTorch version is 1.6.0 and torchvision version is 0.7.1. The cuda version used is 10.0.

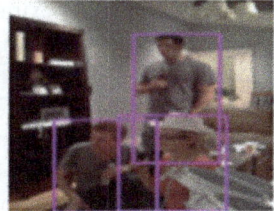

Fig. 1 Detection images

2.2 Object Detection

Moving object detection is mainly to detect moving human targets in video or images and segment them. Whether the location of the human body can be accurately determined will affect the detection rate of fall detection.

Introduction to Yolov5 Algorithm. This paper selects yolov5 as the method of moving human target detection. The basic idea of yolo series of algorithm is to use an end-to-end convolutional neural network to directly predict the category and location of the target, yolov5 has four network structures, yolov5s, yolov5m, yolov5l, and yolov5x; yolov5s is the smallest depth and width of the feature maps. The other three networks are continuously deepened on its basis. The network structure used in this article is the yolov5l.

Experiment and Result Analysis. This paper first uses the yolov5 algorithm to detect the moving human target of the video images. Yolov5 algorithm can detect many types, but this paper only needs to detect the category of person, so this paper changes its detection category to only detect the category of person, and the experiment proves that it has a good detection effect. It not only has a high detection rate for a single person in the background of a simple environment, and it can carry out real-time fall detection, but also has a high accuracy rate of detection in the background of multiple people in a complex environment. The images selected in this paper are from the images in the COCO dataset [11] (website: https://cocodataset. org/), as well as a URL [12] and UR Fall Dataset [13], as well as a URL [14] (website: http://fenix.univ.rzeszow.pl/~mkepski/ds/uf.html). A detection effect is shown in the Fig. 1.

2.3 Fall Detection

This paper uses yolov5 to detect moving human targets, and we have determined moving human bodies in video images. In traditional human fall detection, after detecting moving human targets, feature extraction is required, and then threshold judgment is used or the extracted features are sent into SVM for classification. However, due to the interference of various external factors, manually extracted features are difficult to accurately and completely extract features that can describe human behavior, and the direct use of CNN method can eliminate the step of human feature extraction and directly conduct classification and judgment. The network structure of CNN selected in this paper is ResNet-50.

Database Image Acquisition. The training sample images adopted in this paper are based on the images intercepted from the video of UR Fall Dataset, an online open Dataset. In consideration of making the types of data samples more comprehensive, diverse and random, and suitable for real life, the training sample images adopted in this paper are taken from the video of UR Fall Dataset, an online open Dataset. In the selection of image samples in the video, we aim at various postures in the video, such as walking, sitting, squatting, bending and lying down. We took samples of different forms of the human body for extraction. In addition, considering the setting of some parameters in the training process later, the image size must be greater than 32*32.

Image Preprocessing. Image analysis, image quality directly affects the design and the effect of the precision of recognition algorithm. As a result of our training samples by online dataset of video frame processing and extraction, which extracts the sample images usually having noise interference, so in order to guarantee the quality of the image does not affect the subsequent ResNet-50 network training, this article adopts the way of image preprocessing of extraction in the video image processing. We use the filter method of image processing, image filter to maintain image detail texture feature under the premise of reducing image noise.

At present, the commonly used image filtering methods mainly include mean filtering, Gaussian filtering, median filtering, bilateral filtering, and so on. We compare the following commonly used filtering methods (see Fig. 2).

After the processing effect comparison, we finally choose the Gaussian filter to process the image.

Resnet-50. In this paper, the selection of ResNet-50 is one of the classical network structures. In the convolutional neural network, with the deepening of the network layer, the network training error and test error will increase. To solve this problem, ResNet was born. It adopted a shortcut connection to make the existing network structure further deepened to hundreds of layers without worrying about the training difficulties.

3   Experiments and Results

In this paper, 1000 images were used to verify our algorithm based on the trained ResNet-50, and the results are shown in Table 1.Indicators of detection are shown in Table 2.

$$\text{Accuracy} = \frac{TP + TN}{TP + TN + FP + FN} \tag{1}$$

(a) Original image   (b) Mean image   (c) Gaussian image

(d) Median image   (e) Bilateral image

Fig. 2 Preprocessed images

Table 1 Detection result

	Detected fall	Detected non-fall
Fall (500)	467	33
Non-fall (500)	28	472

Table 2 Detection index

Accuracy	Precision	Recall	F1-score
93.9%	94.3%	93.4%	93.9%

$$\text{Precision} = \frac{\text{TP}}{\text{TP} + \text{FP}} \tag{2}$$

$$\text{Recall} = \frac{\text{TP}}{\text{TP} + \text{FN}} \tag{3}$$

$$\text{FI} = \frac{2 \cdot \text{Recall} \cdot \text{Accuracy}}{\text{Recall} + \text{Accuracy}} \tag{4}$$

where TP stands for predicting fall as fall; TN stands for predicting non-fall as non-fall; FP stands for predicting non-fall as fall; FN stands for predicting falls as non-falls.

The algorithm proposed in this paper can achieve an accuracy rate of 93.9% in the UR Fall Datasets, the online open dataset. It is found that most of the misdetected images are in the form of lying down and falling down or similar forms. In the subsequent studies, we may consider increasing the data amount of training samples to improve the accuracy.

4 Conclusion

The yolov5 algorithm combined with ResNet-50 to realize real-time moving target detection algorithm is proposed. Its application in human fall detection, the algorithm mainly through yolov5 to test the video images of human movement, the detected in the human body image preprocessing, and then processed images into trained ResNet-50 in the network are classified to determine whether a human fall behavior occurs. Meanwhile, the data samples for convolutional neural network training are images from the UR Fall Dataset. The proposed algorithm can achieve 93.9% accuracy rate, and also satisfy the application in actual living condition.

Acknowledgements This work is supported by Hainan Provincial Natural Science Foundation of China (No. 420CXTD439), and the National Science Foundation of China (No. 61661038).

References

1. Mathie, M.J., Coster, A.C.F., Lovell, N.H., Celler, B.G.: Accelerometry: providing an integrated, practical method for long-term, ambulatory monitoring of human movement. Physiol. Measur. **25**(2), R1-20 (2004)
2. Bianchi, F., Redmond, S.J., Narayanan, M.R., Cerutti, S., Lovell, N.H.: Barometric pressure and triaxial accelerometry-based falls event detection. IEEE Trans. Neural Syst. Rehabil. Eng. **18**(6), 619–627 (2010)
3. Tamura, T., Yoshimura, T., Sekine, M., Uchida, M., Tanaka, O.: A wearable airbag to prevent fall injuries. IEEE Trans. Inf. Technol. Biomed. **13**(6), 910–914 (2009)
4. Nguyen, T.T., Cho, M.C., Lee, T.S.: Automatic fall detection using wearable biomedical signal measurement terminal. In: 2009 Annual International Conference of the IEEE Engineering in Medicine and Biology Society, pp. 5203–5206. IEEE, (2009)
5. Chua, J.L., Chang, Y.C., Lim, W.K.: A simple vision-based fall detection technique for indoor video surveillance. SIViP **9**(3), 623–633 (2015)
6. Agrawal, S.C., Tripathi, R.K., Jalal, A.S.: Human-fall detection from an indoor video surveillance. In: 2017 8th International Conference on Computing, Communication and Networking Technologies, pp. 1–5. IEEE, (2017)
7. Liu, H., Guo, Y.: A vision-based fall detection algorithm of human in indoor environment. In: Second International Conference on Photonics and Optical Engineering, p. 10256: 1025644-1-1025644-6. SPIE (2017)
8. Yao, C., Hu, J., Min, W.: A novel real-time fall detection method based on head segmentation and convolutional neural network. J. Real-Time Image Proc. **17**(4), 1939–1949 (2020)
9. Burkhardt, J.H., Lay, C.M.: Fall detection from human shape and motion history using video surveillance. In: International Conference on Advanced Information Networking & Applications Workshops, pp. v–xv. IEEE (2007)

10. He, K., Zhang, X., Ren, S.: Deep residual learning for image recognition. In: IEEE Conference on Computer Vision and Pattern Recognition, pp. 770–778. IEEE (2016)
11. Lin, T.Y., Maire, M., Belongie, S.: Microsoft COCO: common objects in context. In: European Conference on Computer Vision, pp. 740–755. Springer (2014)
12. COCO Dataset Homepage. https://cocodataset.org/
13. Kwolek, B., Kepski, M.: Human fall detection on embedded platform using depth maps and wireless accelerometer. Comput. Methods Programs Biomed. **117**(3), 489–501 (2014)
14. UR Fall Dataset Homepage. http://fenix.univ.rzeszow.pl/~mkepski/ds/uf.html

Author Index

Lightning Source UK Ltd.
Milton Keynes UK
UKHW022054070922
408416UK00001B/2